THE PRODUCTION AND USE OF VEGETABLE OILS IN PTOLEMAIC EGYPT

BULLETIN OF THE
AMERICAN SOCIETY OF PAPYROLOGISTS

SUPPLEMENTS

edited by
Ludwig Koenen (Chair), Ann Hanson,
and Michael Haslam

Number 6

THE PRODUCTION AND USE OF VEGETABLE OILS IN PTOLEMAIC EGYPT

by
D. Brent Sandy

THE PRODUCTION AND USE OF VEGETABLE OILS IN PTOLEMAIC EGYPT

D. BRENT SANDY

Scholars Press
Atlanta, Georgia

THE PRODUCTION AND USE OF VEGETABLE OILS IN PTOLEMAIC EGYPT

D. Brent Sandy

Library of Congress Cataloging in Publication Data

Sandy, D. Brent.
The production and use of vegetable oils in Ptolemaic Egypt.
(Bulletin of the American Society of Papyrologists.
Supplements ; no. 6)
1. Oil industries--Egypt--History. I. Title.
II. Series.
TP680.S26 1989 664'.3'0932 86-22011
ISBN 1-55540-075-2 (pbk. : alk. paper)

Printed in the United States of America
on acid-free paper

Contents

List of Tables

Preface

This monograph brings together what can be known about the oils in use in Ptolemaic Egypt. The information comes primarily from the papyri, though it is supplemented throughout with literary sources, modern science and technology, and oil production in third world countries.

All the oils in use in antiquity were vegetable oils, that is, the oils were extracted from the seeds of various plants, and in the case of olive oil, from the fruit of a plant. Though some oil could be collected from plants that grew wild, the demand for oil meant that oil crops were common in ancient agriculture.

The standard material on oils was previously collected in *Die Landwirtschaft in hellenistischen Ägypten* by M. Schnebel (Munich, 1925). The present essay enlarges and updates that part of Schnebel's helpful book and presents a fresh analysis of all aspects of the oils of Ptolemaic Egypt, a discussion which will enhance the understanding of the documents and culture of ancient Egypt.

The evidence collected in this study includes the material compiled for a dissertation at Duke University (1977) but supersedes that in significant ways: papyri published through 1985 are included here; additional subjects are treated; many secondary sources have been added; the Demotic evidence has been examined; and many conclusions have been modified if not reversed.

I owe thanks to many people for their assistance with this project; to those who have helped shape this study: William Willis, John Oates, Kent Rigsby, Roger Bagnall, Ludwig Koenen, Ann Hanson, and Jim Keenan; and those who have helped with typing, keyboarding, drawing, and editing: Dave Zapf, Bonnie Bowley, Mary Dyck, Emiline Secaur, and members of my family (my wife, Cheryl, my mother, Omega, and my aunt, Fern). To the last I am especially grateful for their love and patience.

Abbreviations

Unless otherwise noted, all dates are B.C. Dates for papyri and other ancient sources appear after the references, often in parentheses. In lists and tables, *a* and *p* represent B.C. and A.D. Shortened titles used for frequently cited sources are indicated in the bibliography by bold-face type.

References to ancient authors:

Dioscorides	περὶ ὕλης ἰατρικῆς in *Pedanii Dioscuridis de Materia Medica*, ed. by M. Wellmann, I, II. Berlin: Weidmann, 1958.
Galen	*Medicorum Graecorum Opera Quae Exstant*, ed. by C. G. Kühn. Lipsiae, 1821–33 (citations are to volume and page number).
Hippocrates	*Oeuvres complètes d'Hippocrate*, ed. by E. Littré. Paris, 1839–61 (citations are to volume and page number).

Other references to ancient authors conform to LSJ and *OLD*.

Measures

arou. (ar.)	aroura
art.	artaba
choi.	choinix
metr.	metretes
ch.	chous
ko.	kotyle
dr.	drachma
ob.	obol

Papyri

Editions of Greek papyri and ostraca are cited according to J. F. Oates, R. S. Bagnall, W. H. Willis, and K. A. Worp, *Checklist of Editions of Greek Papyri and Ostraca, Bulletin of the American Society of Papyrologists*, Supplement 4 (Scholars Press, 1985). In addition, note the following:

O.Fawâkhir — "Ostraca grecs et latins de l'Wâdi Fawâkhir," ed. O. Guéraud, *Bulletin de l'Institut français d'archéologie orientale* XLI (1942) 141–96.

O.Fay — *Fayum Towns and Their Papyri*, B. P. Grenfell, A. S. Hunt, and David G. Hogarth (London: Egypt Exploration Fund, 1900).

Reference works

BDB — Brown, F.; Driver, S. R.; and Briggs, C. A. *A Hebrew and English Lexicon of the Old Testament Based on the Lexicon of W. Gesenius* (Oxford, Clarendon, 1907).

DarSag — Daremberg, C., and Saglio, E. *Dictionaire des antiquités grecques et romaines d'après les textes et les monuments* (Paris: Librairie Hachette, 1877–1919).

EB[11], *EB*[14], *EB*[15] — *Encyclopaedia Britannica*, 11th ed. (Cambridge: University Press, 1910); 14th ed. (London: William Benton, 1971); *The New Encyclopaedia: Micropaedia, Macropaedia*, 15th ed. (London: William Benton, 1974).

LÄ — Helck, W., and Westendorf, W. *Lexikon der Ägyptologie* (Wiesbaden: Harrassowitz, 1972–).

LSJ — Liddell, H. G.; Scott, R.; and Jones, H. S. *A Greek-English Lexicon*, 9th ed., 1940 with a Supplement, 1968 (Oxford: Clarendon).

OCD — Hammond, N. G. L., and Scullard, H. H. *The Oxford Classical Dictionary*, 2nd ed. (Oxford: Clarendon, 1970).

OGIS — Dittenberger, W. *Orientis Graeci Inscriptiones Selectae* I, II (Leipzig: S. Hirzel, 1903–05).

OLD — Glare, P. G. W., ed. *Oxford Latin Dictionary* (Oxford: Clarendon, 1982).

RE — Wissowa, G. *Paulys Real-Encyclopädie der classischen Altertumswissenschaft* (Stuttgart: J. B. Metzler, 1893–1967).

SEHHW — Rostovtzeff, M. *The Social and Economic History of the Hellenistic World* I–III (Oxford: Clarendon, 1941).

TGL — Estienne, H. *Thesaurus graecae linguae.* (Paris: Didot, 1831–65).

Demotic papyri

See the list by Zauzich in *LÄ* (Band IV, Lieferung 5,6; s.v. Papyri, Demotische):

I.dem.Nubie — Griffith, F. Ll. *Les temples immerges de la Nubie: Catalogue of the Demotic Graffiti of the Dodecaschoenus.* Service des antiquités de l'Égypte. Oxford: University Press, 1937.

O.dem.Leiden — Nur El-Din, M. A. A. *The Demotic Ostraca in the National Museum of Antiquities at Leiden.* Leiden, 1974

O.dem.Medinet Habu — Lichtheim, M. *Demotic Ostraca from Medinet Habu.* Oriental Institute Publ. LXXX. Chicago, 1957.

O.dem.Ossirinco — Bresciani, E. "Ostraka demotici da Ossirinco." *Studi classici e Orientale* 19–20 (1970–71):357–99.

O.dem.Theb. — Gardiner, A. H.; Thompson, H.; and Milne, J. G. *Theban Ostraca.* University of Toronto Studies, Philological Series I. London, 1913.

O.dem.Zürich — Wångstedt, S. V. *Die demotischen Ostraka der Universität zu Zürich.* Uppsala, 1965.

P.dem.Adler — Griffith, F. Ll. *The Adler Papyri.* Oxford, 1939.

P.dem.Berlin — Lüddeckens, E., and Kaplony-Heckel, U. *Ägyptische Handschriften,* 2 vols. Wiesbaden: Franz Steiner, 1971.

P.dem.Cairo — Spiegelberg, W. *Die demotischen Denkmäler*; II: *Die demotischen Papyrus.* Strassburg: Schauberg, 1908. III: *Demotische Inschriften und Papyri.* Berlin, 1932.

P.dem.Eheverträge — Lüddeckens, E. *Ägyptische Eheverträge.* Ägyptologische Abhandlungen I. Wiesbaden: Harrasowitz, 1960.

P.dem.Köln — ———. *Demotische und koptische Texte.* Papyrologica Colonensia II. Köln, 1968.

P.dem.Lille — de Cenival, F. *Cautionnements demotiques du début de l'époque ptolémaïque.* Paris, 1973.

P.dem.Loeb — Spiegelberg, W. *Die demotischen Papyri Loeb.* München, 1931

P.dem.Ryl. — Griffith, F. Ll. *Catalogue of the Demotic Papyri in the John Rylands Library,* III. Manchester, 1909.

P.dem.Tempeleide — Kaplony-Heckel, U. *Die demotischen Tempeleide,* 2 vols. Ägyptologische Abhandlungen 6. Wiesbaden, 1963.

P.dem.Zen. — Spiegelberg, W. *Die demotischen Urkunden des Zenon-Archivs.* Leipzig, 1929.

P.London-Leiden — Griffith, F. Ll., and Thompson, H. *The Demotic Magical Papyrus of London and Leiden.* 3 vols. London, 1904–09.

Pap.Lugd.Bat. XX — Pestman, P. W. *Greek and Demotic Texts from the Zenon Archive.* Papyrologica Lugduno-Batava XX. Leiden, 1980.

P.Tebt.Tait	Tait, W. J. *Papyri from Tebtunis in Egyptian and Greek.* Egypt Exploration Society, *Texts from Excavations* 3. London, 1977.
P.Tor.Amenothes	Pestman, P. W. *L'Archivio di Amenothes, figlio di Horos.* Catalago del Museo Egizio di Torino. Serie prima. Monumenti e Testi V. Milano, 1981.
P.Vindob.D. 6257	Reymond, E. A. E. *A Medical Book from Crocodilopolis.* Österreichische National-bibliothek. Vienna, 1976.

I

Introduction

Oil has been a long-standing staple across many cultures and centuries, touching almost every aspect of ancient life in some way. In medicine, oils brought healing to numerous maladies; in cosmetics, oil provided a cleanser and beautifier; in households and businesses, oil fueled the lamps of all classes of society; in agriculture, the sources of oil provided crops for the farmers; in food preparation, oil added flavor and viscosity; in religion, oil was a sacred and mystic substance; in art, the olive tree and branch were motifs of decoration, both symbolic and aesthetic; in death, oil was a regular ingredient in embalming procedures. Thus Pliny *Nat.* 14.29(150) extols the virtues of two liquids, wine and oil; but the latter he considers a necessity.

Oil is especially prominent in religious literature. In Assyrian and Babylonian texts, oil was frequently assigned magical qualities, to the extent that memorial stones were anointed with oil. In the Jewish Scriptures, kings were ceremonially anointed with oil, the tabernacle was consecrated with oil, and in the Talmud, specific instructions were given on the use of oils. For Christians, the Mount of Olives and the adjacent Gethsemane, which in Hebrew means oil press, became especially sacred sites.

Though olive oil has been the most common oil in the Mediterranean world, both in antiquity and in modern times, it is not alone. Today, vegetable oils are derived from over forty species of trees and plants, and in ancient times, the sources were nearly as many. Alexander Trallianus mentions twenty-nine different types of oil in his medical prescriptions, a list that does not include sesame, safflower, and linseed.[1] Pliny *Nat.* 15.7(24), however, considered all oils except olive oil to be artificial (*ficticium oleum*).

[1] See the index in T. Puschmann, *Alexander von Tralles Original Text und Übersetzung, ein Beitrag zur Geschichte der Medicin* (Vienna, 1878), Vol. II., p. 605.

§ 1 *The Evidence*

The Revenue Laws[2] represents the longest surviving Ptolemaic papyrus and is as well the single most informative text about oils, although there are a variety of papyri that mention oils more briefly. The full but sometimes unclear details of ***P.Rev.*** include every stage of the production of oil (these details regarding oils occupy columns 38–72 of the 107 columns extant). Prices and tariffs are stated for different oils and oil seeds; regulations are given for harvest, for oil factories, and for transporting oil; penalties are prescribed for any who disregard the oil law; individual nomes are assigned amounts of land to be planted in oil seeds, etc.

P.Rev. has been fundamental to modern understanding of Ptolemaic oils and has provided the basis of important research on the Ptolemaic economy.[3] Bingen, who wrote a commentary on ***P.Rev.*** for his doctoral dissertation (never published; it is in the Fondation in Brussels), has written an important monograph[4] on ***P.Rev.*** and the Ptolemaic economy. Whereas others were mesmerized by ***P.Rev.***, Bingen has offered a reappraisal that sees ***P.Rev.*** as at least seven separate documents, each a fiscal measure to meet the immediate problems faced by Philadelphus' administration. Bingen concludes that ***P.Rev.*** had little effect on agricultural production and did not determine the direction of the Ptolemaic economy.[5]

[2] Subsequent references to the Revenue Laws are abbreviated ***P. Rev.*** The editions of the text are: B. P. Grenfell and J. P. Mahaffy, *The Revenue Laws of Ptolemy Philadelphus* (Oxford: Clarendon Press, 1896); Ulrich Wilcken, *Grundzüge und Chrestomathie der Papyruskunde;* I. Band, *Historischer Teil;* II. Hälfte, *Chrestomathie* (Leipzig: Teubner, 1912): No. 258 = ***P.Rev.*** 1–22; No. 249 = ***P.Rev.*** 36–37; No. 299 = ***P.Rev.*** 38–58; and No. 181 = ***P.Rev.*** 73–78; Jean Bingen, *Papyrus Revenue Laws,* Beiheft I, *Sammelbuch griechischer Urkunden aus Ägypten* (Göttingen: Hubert and Company, 1952); this edition includes introduction, bibliography, and critical apparatus. Portions of ***P.Rev.*** have also been published in: A. S. Hunt and C. C. Edgar, *Select Papyri with an English Translation,* II. *Official Documents,* The Loeb Classical Library (Cambridge: Harvard, 1934), No. 203 = ***P.Rev.*** 38–53.3; 54.15–56.18 (the basis of this text is Grenfell's editio princeps with some revisions); and M. David and B. A. van Groningen, *Papyrological Primer* (Leiden: E. J. Brill, 1940), No. 17 = ***P.Rev.*** 24.14–25.16.

[3] Rostovtzeff, *SEHHW, Large Estate,* and "Foundations"; Préaux, *L'économie,* and "Sur les origines des monopoles Lagides," *CE* 29:57 (1954) 316–27. Brief discussions of ***P.Rev.*** are in: Wilcken, *Grundzüge und Chrestomathie der Papyruskunde* I.I. pp. 239–48, and Heichelheim, "Monopole," *RE* XVI.1 (1933): 165–70.

[4] Bingen, *Le Papyrus Revenue Laws—Tradition grecque et adaptation hellénistique,* Rheinisch-Westfälische Akademie der Wissenschaften, Vorträge, G 231 (Opladen: Westdeutscher, 1978). Cf. Bingen, "Les colonnes 60–72," *CE* 41 (1946) 127–48 and "Économie grecque et société égyptienne au IIIe siècle," *Das ptolemäische Ägypten,* ed. by H. Maehler and V. M. Strocka (Mainz am Rhein: Philipp von Zabern, 1978).

[5] Bingen, *Le Papyrus Revenue Laws,* p. 19ff.

Considerable amounts of data about Ptolemaic oils come from numerous other papyri; in particular, two archives figure prominently: the Zenon collection[6] frequently mentions oil seeds and land planted in oil crops; the texts from the Serapeum at Memphis[7] are especially concerned with daily quantities of oil. Demotic papyri[8] also demonstrate the importance of oils to the native population.

In addition to the papyrological evidence, the literary records of Theophrastus, Dioscorides, Pliny, and others help to clarify identifications of the plants cultivated for their oily seeds. For an understanding of agricultural implements and oil factories, archaeological data is also applicable.

§ 2 *Sources of Oil*

After the Greek takeover of Egypt, castor seed, sesame seed, olives, and safflower seed were processed for their oil content. Castor oil was adequate for lighting, but its odor and purgative qualities made it less attractive for food preparation. Sesame was a good oil for lamps and was also very useful in foods. A limited supply of olives prevented olive oil from dominating oil consumption in Ptolemaic Egypt. The same was true of safflower. Castor, sesame, and safflower are today noted for their common characteristics: the plants are adapted to a wide range of climatic conditions, the seeds and oils can be stored easily and for extended periods of time, and the oils have many uses.

Evidence for other oil sources in Ptolemaic Egypt is slight. ***P.Tebt.*** III pt. II 829 (180–79? B.C.), a land survey, has a heading, "oil crops," and then lists the number of arouras planted in safflower, castor, vetch (ἀ̣ρ̣ά̣κ̣ω[ι]), and hay (χόρτος).[9] The possibility of processing hay for its oil is confirmed by Pliny *Nat.* 15.7(30); he reports that a large amount of oil is obtained in Egypt from *chortinus*.[10] Hay was a common crop in Ptolemaic Egypt,[11] but texts that indicate its use reveal that it was generally fodder for animals. Other than Pliny and the papyrus cited above, no ancient evidence has been found for oil produced from grass or vetch. Modern studies report that the seed of clover, alfalfa, etc., can be pressed for its oil, though the seed of such grasses only contains about 6–8% oil.[12]

[6] See P. W. Pestman, *A Guide to the Zenon Archive*, ***Pap.Lugd.Bat.*** XXI A, B.

[7] ***UPZ*** I.

[8] Below §7, pp. 30–34.

[9] Cf. ***PSI*** IV 432.2–3, a cultivator's letter to Zenon asking whether to plant vetch, sesame, or hay; also, ***P.Petr.*** II 38a.23 mentions hay at a threshing floor.

[10] Galen 19.725 allows sesame seed to be substituted for vetch seed.

[11] Schnebel, *Landwirtschaft*, pp. 211–18.

[12] Eckey, *Fats and Oils*, p. 485.

P.Rev. suggests two oil sources in addition to sesame, castor, and safflower.[13] The oil section begins by giving the prices of the oil seeds in descending order:

sesame	8 dr.
castor	4 dr.
safflower	1 dr. 2 ob.
gourd[14]	4 ob.
linseed[15]	3 ob.

Next, the price of the oil itself is given. Though it was corrected to read 48 dr. for each oil, it originally read 48 dr. for sesame and safflower, and 30 dr. for castor, gourd, and lamp oil (ἐπελλύχνιον has replaced linseed). Thereafter, sesame and castor are frequently mentioned while the other three are mentioned infrequently:

safflower	16 times
gourd	9 times
linseed	3 times (+ ἐπελλύχνιον 3 times)

Only sesame and castor are specified in cols. 60–72 for crop production in the nomes.

Nowhere else in the papyri or in any ancient source is gourd seed oil[16] or linseed[17] oil attested. Perhaps ***P.Rev.*** regulated gourd seed and linseed because every precaution was being taken to protect

[13] ***P. Rev.*** does not mention olives as a source of oil, though it does regulate importing oil (see below §6, pp. 24–25).

[14] The identification of κολόκυντος is not secure. Pliny *Nat.* 20.8(14) classes it with *cucumis* and *cucurbita;* Dioscorides 4.176(178) says that the leaves spread on the ground, that the fruit is round and bitter, and that it must be picked when it turns pale. In modern literature, κολόκυντος has been identified with *Citrullus colocynthis,* a melon, and with *Cucurbita maxima L.,* a squash. The latter, however, was apparently not grown in Europe before the discovery of the new world. Until more precisely identified, κολόκυντος is best translated gourd, a term which designates any of the *Cucurbitaceae* including various types of melon, squash, and cucumber. Schnebel, *Landwirtschaft,* pp. 202–03; LSJ p. 973; *EB*[15] 9.502.

[15] *Linum usitatissimum L.* (Fam. Linaceae) is still used to produce cloth and is now widely cultivated to produce linseed oil. Pliny *Nat.* 1-6(1-25) gives an extensive description of the flax plant and its characteristics. Theophrastus *HP* 3.18.3 describes the stickiness of the seeds. Galen 19.742 says that linseed may be substituted for sesame seed in medical prescriptions.

[16] The occurrences of κολόκυντος (including variant spellings, see Mayser, *Grammatik*) are: 3rd cent.: ***P.Cair.Zen.*** I 59033.14; II 59292.132, 319; III 59300.3; V 59838.6; ***P.Lille*** I 58ii.15; ***P.Lond.*** VII 2172.26; ***P.Mich.*** I 50.2; ***P.Rev.*** 39.6; 40.10, 12; 53.22; 55.6, 9; 57.16 (= 59.19); 57.18 (= 59.21); 58.2 (= 60.5); ***PSI*** IV 402.5, 11; 434.3; VI 553.14.

the oil industry and the government's interests. In the event someone thought he could obtain oil from gourd seed or linseed and not be subject to governmental control, the oil law was written to include gourd and linseed in its provisions. Apparently for the same reasons, animal fat was regulated lest anyone melt that to obtain oil.[18] Without other evidence to support the references to gourd seed oil and linseed oil in ***P.Rev.***, it cannot be assumed that these oils were in use in Ptolemaic Egypt.

Linseed oil is today the most important of the drying oils.[19] More linseed oil is currently produced than sesame, castor, or safflower. And attention has recently been drawn to the seeds of the gourd family (*Cucurbita* species) as a valuable domestic source of oil.[20]

Several oils were in common use before and after Ptolemaic Egypt: ben and balanos oils[21] were widely used in Pharaonic Egypt;

2nd cent.: ***P.Tebt.*** I 131 (see *ZPE* 41 (1981), pp. 263–69); III pt. II 1093.6. 1st cent.: ***BGU*** XIV 2449.66(?). Roman: ***BGU*** IV 1120.13; ***P.Berl.Leihg.*** 18.15; ***P.Hamb.*** I 99.8; ***P.IFAO*** II 7.8; III 14.7; 37. *passim*; ***P.Lond.*** V 1881.3; ***P.Mich.*** II 123v(6).36; (7).26; ***P.Oxy.*** XXIV 2423r(ii).23; ***P.Princ.*** II 39.4, 8; ***P.Ross.Georg.*** II 19.14; 41.5; ***P.Ryl.*** IV 629. *passim;* 630. *passim;* ***PSI*** XIII 1338.9; ***P.Wash.Univ.*** 52.12; ***SB*** VI 9017 xiii.9 9408(2).87; 9409(7). 96; ***O.Fawâkhir*** 13.9. Gourds were grown in Egypt before the Ptolemaic period, but there is no evidence that the seeds were pressed for oil. Woenig, *Pflanzen*, pp. 203–4; Keimer, *Gartenpflanzen*, p. 13.

[17] The occurrences of λίνου σπέρμα are: 3rd cent.: ***BGU*** VII 1523.1, 4–5; ***P.Cair.Zen.*** II 59292.*passim;* IV 59730 (see ***P.Lond.*** VII 1991); ***P.Lond.*** VII 1991.79, 105; 1994.*passim;* 1995.*passim;* ***P.Mich.*** I 26.4; ***P.Rev.*** 39.7; 55.6; 57.19 (= 59.22); ***P.Tebt.*** III pt. I 769.13, 46, 48, 78. 1st cent.: ***BGU*** XIV 2449.25, 58, 78, 130. Roman: ***BGU*** XIII 2327.14, 16; ***P.Coll.Youtie*** II 80.17 (= ***P.Oxy.*** XLV 3255); ***P.Mich.*** VI 521.17; ***P.Mil.Vogl.*** VI 276.11; ***P.Oxy.*** I 103.13; ***PSI*** VIII 959.18. This list does not include occurrences of λίνον; since this plant was used to produce cloth—especially in Egypt both before, during, and after the Ptolemies—references to λίνον probably have no bearing on the possible production of linseed oil. Cf. F. Olck, "Flachs," *RE* VI (1909), cols. 2435–84; Woenig, *Pflanzen*, pp. 181–89; Rostovtzeff, *SEHHW*, p. 377; J. Grant, "A Note on the Materials of Ancient Textiles and Baskets," *From Early Times to the Fall of Ancient Empires*, vol. I of *A History of Technology* (Oxford: Clarendon, 1954), p. 448; R. Patterson, "Spinning and Weaving," *The Mediterranean Civilizations and the Middle Ages*, vol. II of *A History of Technology* (Oxford: Clarendon, 1956), p. 195.

[18] ***P.Rev.*** 50.14–19 makes the sale of, storing of, or melting of animal fat illegal. Textual and archaeological evidence is plentiful for the use of animal fats in Pharaonic Egypt (Lucas, *Egyptian Materials*, pp. 327–28, 330), but for Ptolemaic Egypt the opposite is true. Butchers are frequently mentioned in the papyri but no references to the processing of fat into oil or of fat used in place of oil occur. The word στέαρ occurs a total of 4 times: ***P.Rev.*** 50.14; ***P.Cair.Zen.*** II 59176.183 (for launching boats), IV 59703.2, 6 (in a list with cheese, milk, and water); and ***P.Col.*** III 43.5 (for a boat).

[19] Vaughan, *Oil Seeds*, p. 141.

[20] Eckey, *Fats and Oils*, pp. 763–64.

[21] Lucas, *Egyptian Materials*, pp. 330–31.

radish seed oil[22] and vegetable seed oil[23] were common in Roman Egypt; but these oils are without evidence in Ptolemaic Egypt.[24] Poppy seed, once thought to be a source of oil, is unattested as an oil source.[25]

The sources of oil treated in this study are therefore four: castor, sesame, olive, and safflower.

§ 3 *Oil Seed Agriculture*

The success of agriculture in Egypt, and hence the prosperity of the entire economy, depended on the Nile River and in particular on its annual flood.[26] Without the Nile, Egypt would be an uninhabitable and infertile desert.

On account of the summer rainy season far to the south of Egypt, the Nile began to rise in June, but the maximum flood level, as much as 15 m. above normal, was not reached until mid-September. By the end of October, however, the river was back within its banks and agricultural work could begin. Some fluctuation was possible both in the extent of the flood and in the dates when it overflowed its

[22] Pliny *Nat.* 15.7(30) and 19.26(79) says that more oil is obtained in Egypt from radish seed than from any other plant. References to radish oil are common in the Roman papyri. And in the Talmud (*Shabbath* 26a) it is reported that only radish oil is available in Alexandria. Radishes were also a crop of Pharaonic Egypt (Woenig, *Pflanzen*, pp. 216–18), but evidence of radish oil is lacking.

[23] Frank, *Economic Survey*, II, p. 3. Numerous Roman papyri mention λαχανόσπερμον; vegetable seed is identified as a source of oil on the basis that it is measured with the μέτρον ἐλαιουργικόν (H. C. Youtie, "Four Short Texts on Papyrus," *ZPE* 29 [1978], p. 287 n. 6). What vegetable is intended by λάχανον is ambiguous (Schnebel, *Landwirtschaft*, p. 210).

[24] The Ptolemaic papyri refer infrequently to radishes and to radish seed; one possible reference to radish oil exists: ***PSI*** V 537.15 ῥαφανε[λαίου?]. On the omission of radish from ***P.Rev.***, see Schnebel, *Landwirtschaft*, p. 208. No occurrences of λαχανόσπερμον exist, though λάχανον is fairly common.

[25] In his commentary on ***P.Edg.*** 103, Edgar suggested that poppy seed was a part of the oil monopoly because he thought that Thrasymedes, to whom Zenon sent sesame and poppy seed, was an agent of the oil monopoly. But after the publication of ***P.Lille*** 39–51 which clarified Thrasymedes's position, Edgar said (on ***P.Cair.Zen.*** II 59223) that his former suggestion is unlikely.

[26] D. Bonneau, "Le Nil à l'epoque ptol.," *Travaux de la maison de l'Orient*, II: *L'homme et l'eau en Mediterranee et au Proche Orient* (Lyon, 1981), pp. 103–14; D. Bonneau, *La crue du Nil, divinité égyptienne à travers mille ans d'histoire* (Paris: Librairie C. Klincksieck, 1964); see also Schnebel, *Landwirtschaft*, pp. 137–82, Johnson, *Roman Egypt*, pp. 7–25, and the next note; for modern Egypt, see H. Lyons in Baedeker, *Egypt and the Sudan*, pp. lxvi–lxix.

banks.[27] Consequently, a few floods have been so significant that famine threatened the population of Egypt.[28]

The Nile at flood stage benefited Egyptian agriculture in important ways: water, normally confined to the river bed and only elevated to the surrounding land with great effort, was widely distributed throughout the Nile valley; this distribution of water provided not only the necessary moisture, but deposited a fertile silt that renewed the land. Because the flood was an annual event, the moisture and fertility of the soil were revived each year.

Irrigation. It was not enough to allow the flooding Nile to come and go as it pleased if the agricultural potential was to be realized. One soaking per year was hardly sufficient moisture for most crops. Thus, elaborate systems of basins to retain the water, canals to transfer the water, drains to remove excess water, and lifting devices to raise the water supply to higher elevations were in use in Ptolemaic Egypt.[29] If these irrigation devices were carefully administered and maintained, the benefits of the Nile provided for an abundant harvest.

For the Fayum, evidence exists in the Zenon archive of extensive work of constructing new irrigation systems. Because of below-sea-level elevation, the Fayum held the promise of growing crops that would not be dependent on the flooding of the Nile since water could be channeled in year-round. Thus, numerous dikes, drainage ditches, and canals were built to bring water to a larger area of the Fayum; at the same time, brushwood and trees were cut in order to reclaim as much land as possible for agricultural crops.[30]

Crop Seasons. On the land subject to the Nile floods, crops were usually planted in October, November, or December after the

[27] For the time of the Nile floods, see D. Bonneau, *Le fisc et le Nil, incidences des irrégularités de la crue du Nil sur la fiscalité foncière dans l'Egypte grecque et romaine,* Publications de l'institut de droit romain de l'université de Paris, n.s. 2 (Paris: Editions Cujas, 1972), p. 9, nn. 2–3.

[28] E.g. while the Jews were in Egypt, Gen 41:25–57; and e.g. in 48 B.C. under Cleopatra, Suetonius *Aug.* 18.2.

[29] On irrigation see ***P.Tebt.*** III pt. I 703.29–40; Rostovtzeff, *SEHHW* I, pp. 274–75; W. Willcocks and J. I. Craig, *Egyptian Irrigation,* 3rd ed. (New York: Spon and Chamberlain, 1913); W. Kosack, *Historisches Kartenwerk Ägyptens* (Bonn, 1971); K. A. Wittfogel, *Oriental Despotism: A Comparative Study of Total Power* (New Haven: Yale, 1957); and Cooper, "Rules for the Ministries," pp. 56–57. On the lifting devices, see Crawford, *Kerkeosiris,* p. 107 n. 3, and Rostovtzeff, *SEHHW,* pp. 363–64.

[30] Rostovtzeff, *Large Estate,* pp. 59–69; on the Fayum, see Johnson, *Roman Egypt,* pp. 10–12, and the excavation reports (on the irrigation system), G. Caton-Thompson and E. W. Gardner, *The Desert Fayum* (Royal Anthropological Institute, 1934), pp. 140–57.

flood receded; harvest then came in the spring.[31] Temperatures throughout the winter months were temperate enough for most crops, though in modern times January may bring cool nights (as low as 8° C) to Cairo.[32] On lands of higher elevation that were not inundated by the flood and that were irrigated manually, and on lands in the Fayum irrigated by canals, crops were not limited to the winter growing season. As will be demonstrated below (ch. 3 §7, pp. 63–64), a two-crop season of sesame was accomplished in some parts of Ptolemaic Egypt, and in the Fayum, wheat was planted twice per year.[33] Crop rotation is also attested for Ptolemaic Egypt; thus, there was some effort to prevent the soil from becoming exhausted by planting different crops from year to year.[34]

Since the construction of several dams on the Nile in this century, Egypt is mostly under perennial irrigation rather than the annual flood. Hence three growing seasons are possible, depending on crops and location.[35] Grains are planted in October and November and harvested in April and May; rice and cotton are planted in the spring and harvested in September and October; a few crops that need only a two-month growing season are planted at the beginning of October and harvested at the end of November. The first of these seasons is still the most commonly utilized; the last accounts for only 17% of the crop totals.

Implements. Though no mention of agricultural implements has been found in the papyri concerning oils, most of the tools used for centuries in the cultivation of Egyptian crops were in all probability also employed in Ptolemaic times for the oil seed plants.[36] Likewise, the tools of Roman farming for which ample evidence exists also suggest the sorts of implements used in Ptolemaic Egypt.[37]

31 Schnebel, *Landwirtschaft*, pp. 138–40, lists 33 papyri that for the most part report a fall planting.

32 H. Lyons in Baedeker, *Egypt and the Sudan*, p. lxxx.

33 ***P.Cair.Zen.*** II 59155.

34 Schnebel, *Landwirtschaft*, pp. 220–28, and Crawford, *Kerkeosiris*, p. 116 n. 6.

35 H. Lyons in Baedeker, *Egypt and the Sudan*, p. lxxv; D. N. Wilber, *United Arab Republic: Egypt, Its People, Its Society , Its Culture* (New Haven, HRAF, 1969), p. 307.

36 Rostovtzeff, *SEHHW* I, p. 364; Schnebel, *Landwirtschaft*, pp. 120–82; for the tools of earlier Egypt, see W. C. Hayes, *The Scepter of Egypt: A Background for the Study of the Egyptian Antiquities in the Metropolitan Museum of Art*, Part II: *The Hyksos Period and the New Kingdom (1675–1080 B.C.)* (Cambridge: Harvard University, 1959), pp. 215–16.

37 K. D. White, *Agricultural Implements of the Roman World* (Cambridge: Cambridge University, 1967); idem., *Farm Equipment of the Roman World* (Cambridge: Cambridge University, 1975).

The more advanced agricultural machines described by Cato (*Agr.*) were, however, apparently not in use in Ptolemaic Egypt, to judge from archaeological evidence.[38] Because the design of Egyptian tools tended not to conform to the designs most common in Europe, implements described in literary sources, which are valid for Italy, may not have been in use in Ptolemaic Egypt.

In general, the tools for oil seed cultivation were a plow for loosening the soil and preparing it for planting, a hoe for loosening and clearing the land for planting, a rake for covering the seed, and a sickle for harvesting the crop.[39] The plow was two or three pieces of wood joined together, one attached to an animal to pull it, one (often iron-tipped) to turn over the soil, and one as a handle for the cultivator. The hoe was usually two pieces of wood in an inverted V-shape, one part the handle and the other the blade for loosening the soil; the two parts were normally held together by rope. The rake was all of wood with notches carefully cut on one side of one piece and a wooden handle attached.

The sickle was a piece of curved wood with a groove of approximately 40 cm. for flint teeth or an iron blade. These primitive yet functional tools continued in use into the early part of the present century in much of Egypt.[40]

Measures. Particularly obscure and confusing are the quantities used in the papyri for seeds and oils. When the Ptolemies arrived in Egypt, they found in place a totally different system of measures than that to which they were accustomed. Typical of the Ptolemaic administrative approach to the native population, the local procedures were in the main not replaced by the Greek system of measurement; on the contrary, the Greek system was adjusted to accommodate the peculiarities of the already complicated Egyptian system. Greek and sometimes Persian terminologies were retained, but on Egyptian standards. Capacities differed according to types of goods, though the same terms were normally used. Thus, there is a frequent lack of uniformity: the capacities known for Greece do not apply to Ptolemaic Egypt, though the terminology is the same; capacities for wheat and wine may or may not be the same as that for sesame seed and castor oil.[41] In spite of the difficulties regarding

[38] W. M. F. Petrie, *Tools and Weapons Illustrated by the Egyptian Collection in University College, London, and 2,000 Outlines from Other Sources* (London: British School of Archaeology in Egypt, 1917), pp. 62–65.

[39] The most helpful discussion is by Petrie, complete with many pictures; see the preceding note.

[40] H. Lyons in Baedeker, *Egypt and the Sudan,* p. lxxv.

[41] Rostovtzeff, *SEHHW* II, pp. 1296–1300.

measures, it is possible to make some determinations, albeit with uncertainty.

The standard discussion of Ptolemaic measures has been Wilcken's,[42] though it has been improved by numerous studies. On the basis of his work and more recent work,[43] the following measures and equivalencies will be used throughout this study of oils:

Area:
1 ἄρουρα = 100 square πήχεις = .23 ha. = .57 acre
1 *iugerum* = 2 square *actus* = .25 ha. = .62 acre

Dry capacity:
1 ἀρτάβη of 40 χοίνικες = 43.12 l. = 1.22 bu.[44]
1 χοῖνιξ = 1.08 l.
1 *modius* of 16 *sextarii* = 8.75 l.

Liquid capacity:
1 μετρητής of 12 χόες = 39.39 l.[45]
1 μετρητής of 6 χόες = 29.55 l.
1 χοῦς = 12 κοτύλαι
1 ἡμικάδιον = ½ μετρητής = 3 χόες[46]

42 U. Wilcken in ***O.Wilck.*** I, pp. 738–80.

43 A. Segrè, *Metrologia e circolazione monetaria degli Antichi* (Bologna: Nicola Zanichelli, 1928), pp. 20–28, 497–508; for corrections of Segrè, see O. M. Pearl, "Varia Papyrologia," *TAPA* 71 (1940) 372–90; very helpful is ***Pap.Lugd.Bat.*** 21b, pp. 548–52; see also Crawford, *Kerkeosiris,* p. 12 n. 2; for Roman measures see especially R. Duncan-Jones, *The Economy of the Roman Empire: Quantitative Studies,* 2nd ed. (Cambridge: Cambridge University, 1982), pp. 369–72, and E. Brehaut, *Cato the Censor: On Farming* (New York: Columbia, 1933), p. xlvi; see also the next note.

44 Several recent articles by Duncan-Jones and Shelton, though at odds on some details, are improving our understanding of the dry measures in the papyri: Duncan-Jones argues for a stable choi. measure and varying sizes of art.; Shelton sees the art. as a reasonably stable quantity and the choi. as 1/40 of the art. R. P. Duncan-Jones, "The Choenix, the Artaba and the Modius," *ZPE* 21 (1976) 43–52; "Variation in Egyptian Grain Measure," *Chiron* 9 (1979) 347–75; J. Shelton, "Artabs and Choenices," *ZPE* 24 (1977) 55–67; "Two Notes on the Artab," *ZPE* 42 (1981) 99–106; cf. J. Jahn, 'Zum Rauminhalt von Artabe und Modis castrensis," *ZPE* 38 (1980) 223–28. Additional bibliography is given in the above articles. Shelton's calculation of the equivalency of the art. is 43.12 l. (*ZPE* 42 [1981] 101).

45 Wilcken reports another calculation that yields 36.47 l., ***O.Wilck.*** I, p. 762; the ch. of the Attic metr. is smaller than the ch. of the Egyptian metr.; the Attic metr. is 1⅓ times larger than the metr. of Egypt (see ***P.Ryl.*** IV 564.16–17).

46 I give the equivalency of the hemikadion as a standard of measure; it could also be a jar in which case it was considered equivalent to ½ κεράμιον = ½ χοῦς (or sometimes ⅜ χοῦς); see ***Pap.Lugd.Bat.*** XXIb, p. 550.

§ 4 *Oil Factories*

The techniques employed in modern oil industry[47] vary according to oil source and country. The variety of techniques, however, involves a series of standard steps. If a harvest has allowed foreign matter such as twigs and leaves to become mixed with the seed, the seed must first be cleaned, usually by machine. Next, the hulls are removed from the kernels in decorticating devices, and then the meats are crushed by heavy iron rollers. After being cooked for at least an hour, the seed is pressed by hydraulic or screw press. The hydraulic press applies pressure to a quantity of the seed pulp at about 350 kg. per cm. and extracts all but 5–8% of the oil. The screw press, a steel worm revolving inside a long steel barrel, applies pressure at about 4,000 kg. per cm. and extracts all but 3–5% of the oil. Sometimes solvent extraction will be used after the hydraulic or screw press. This involves soaking the remaining seed cake with a solvent that dissolves the oil; after extraction the solvent is evaporated out to leave only oil. By this method virtually 100% of the oil available in vegetable seed is extracted. The final step is refining. Depending on the impurities and moisture present in the oil, any of several processes may be used. The oil may be treated with alkali; or it may be agitated with water, separated by centrifuge, and then sprayed into a vacuum dryer. These processes produce a dry, pure, and bleached oil suitable for food or industrial products.

Before modern advances in technology, oil seeds were commercially processed largely by manual labor, with stone crushers driven by animals, and with lever presses. In some countries these methods are occasionally still in use.[48]

Literary Evidence. From several sources we learn much about the ancient methods of making oil. Regarding castor, we have the following descriptions:

Herodotus Va
2.94 Castor seed is processed in two ways: it is either pounded and then pressed, or it is boiled and the oil from it is collected.

Dioscorides Ip
1.32(38) Castor seed must first be spread out to dry in the sun until the shells break off. After it is pounded in a mortar, it is boiled in water until the moisture has

[47] Emory/Wolf, *Study of Practices,* pp. 13, 28; Weiss, *Castor, Sesame, and Safflower,* pp. 761–92.

[48] Weiss, *Castor, Sesame, and Safflower,* pp. 761–66.

boiled away. The oil can then be skimmed off and stored. The Egyptian method of preparation involves cleaning it, grinding it in a mill, putting it in bags, and squeezing it in a press.

4.161(164) After castor seed is hulled, it is pressed to obtain oil.

Pliny *Nat.* Ip

15.7(25) Castor seed is boiled in water and the oil is collected. In Egypt it is salted and pressed.

The terminology for the processing of castor as it appears in the above citations is as follows:

cleaning	καθαίρω
hulling	λεπίζω, ἀπορρήγνυμι
pounding	κόπτω
mortar	ὅλμος
grinding	ἀλήθω
mill	μύλη
pressing	ἀπιπόω, ἀποθλίβω, ἐκθλίβω, *exprimo*
press	ὄργανον

Only one description of processing sesame is extant:

Pliny *Nat.* Ip

18.23(98) Sesame seed is first soaked in hot water, next it is rubbed to loosen the hull, and then it is placed in cold water so that the chaff floats. Thereafter it must be quickly dried in the sun to prevent mold.

From Mesopotamian sources more can be learned about producing sesame oil. Sesame seeds were first mashed, then either pressed to obtain the oil, or placed in hot water so that the oil would float to the surface.[49] And in the Talmud (*Baba Bathra* 25b, 68a), it is reported that sesame seeds were spread out on a roof to dry, and that some sesame oil makers shook a neighbor's villa when they crushed the sesame seeds!

For olive oil production, there are several detailed descriptions.[50] Olives were first cleaned and then dried for no more than three days. Next they were softened either by soaking in water or by a slight

[49] Levey, *Ancient Mesopotamia,* pp. 90–91.

[50] Cato *Agr.* 18–22, 64–69 (III–IIa); Pliny *Nat.* 15.6(20–23) (Ip); Columella 12.52.2–54.2 (Ip).

pressing. Then they were crushed in one of four different kinds of mills: *mola olearia, trapetum, solea et canalis, tudicula.* The *trapetum,* for example, was a dual action heavy stone mortar; two large convex millstones (*orbes*) turned on a long wooden axle (*cupa*) inside a stone base (*mortarium*). Finally the crushed olives were pressed in a lever press or lever and screw press. The press consisted of a lever (*prelum*) anchored at one end; under this end the container of olive pulp was placed. The other end was then forced down to apply pressure to the pulp.

Pliny *Nat.* 18.74(317) describes the changes in design of presses in the previous 100 years;[51] the old style used ropes and leather straps to apply pressure to the lever, but more recent inventions used an upright beam with spiral grooves (screw) that applied the necessary pressure as it was turned.

Based on these descriptions, especially the exact detail of construction given by Cato, an oil factory's equipment and manner of operation can be reconstructed.[52] To what extent the information from these later sources is informative on Ptolemaic Egypt must, however, be controlled by other evidence.

Archaeological Evidence. Of the *trapetum,* good examples have been found at Stabiae, Boscoreale, Oliaro, and Casa di Miri. For the *mola olearia,* the Rondanini relief provides a clear representation.[53] This is, however, the extent of the evidence, and for the rest of the ancient world, no examples of complete presses have been found. Only press-beds, mortars, and mills are attested.[54]

Papyrological Evidence. Selected Roman papyri provide significant information about oil presses.[55] The terminology in these

[51] The design of oil presses and wine presses was nearly the same; Cato *Agr.* 19 gives the differences in dimensions for the grape and olive presses.

[52] A. G. Drachmann, "Ancient Oil Mills and Presses," *Det Kgl. Danske Videnskabernes Selskab., Archaeologisk-kunsthistoriske Meddelelser,* I.1 (Copenhagen: Levin and Munksgaard, 1932); E. Besnier, *DarSag,* IV, pp. 166–69, "Olea, Oleum"; K. D. White, *Farm Equipment of the Roman World* (London: Cambridge Univ., 1975), pp. 225–33; cf. L. A. Moritz, *Grain Mills and Flour in Classical Antiquity* (Oxford: Clarendon, 1958). White's discussion is for the most part correct, but he wrongly thinks that the olive pits were separated from the pulp before pressing. As noted in ch. 4 §1, p. 73, the oil in the pit is so nearly like the oil in the pulp that it is unnecessary to remove the pits. Columella's comment (12.52.6) does not suggest that the pits were removed, but only that he prefers that they not be crushed. Drachmann saw it correctly; the pits were left in the pulp when pressed.

[53] Drachmann, see preceding note.

[54] Moritz, see note 52 above; and W. R. Patton and J. L. Myres, "On Some Carian and Hellenic Oil Presses," *JHS* 18 (1898) 209–17; they note numerous examples of the stone trough that collected the oil from the pressings. See below, p. 82, n. 18.

[55] ***Chrest.Wilck.*** 176; 312; ***P. Sorb.*** inv. 2371; ***P.Fay.*** 95; ***Stud.Pal.*** XXII 177; ***P.Oxy.*** LI 3639. For a partial list of oil presses, see Johnson, *Roman Egypt*, pp. 364–69. Cf.

papyri, though sometimes not securely identified, provides a picture of the oil pressing equipment in use:

oil factory	ἐλαιούργιον
mortar	θυεία
press	ὄργανον, μηχανή (ἐλαιουργική)
equipment	ἐπισκευή
axle	ἄξων
axle-box	χοινικίς
beam	δοκός, ὦμος (?)
prop	ἔρεισμα
tool	ξυλικὸν ἀργαλεῖον
kettle	χαλκίον
worker that rubs	τριβεύς
mill stone (?)	στρόβιλος
mill stone (?)	κάλαθος

Though it may not be possible to match up all the parts of the oil factories in Roman Egypt with the descriptions in literary sources, the similarity of equipment is apparent. (***PSI*** IX 1030 reveals that an olive factory could be part of a private house and that its equipment was made from wood and stone.)

P.Rev. is the most informative Ptolemaic papyrus on processing oils. The regulations on oil factories and oil workers are numerous: each mortar was required to process one artaba of sesame per day, four of castor, or one of safflower; each factory was to be registered with the government, and when not making oil, the equipment was to be locked up; oil workers were not allowed to move from nome to nome; etc. These regulations are given principally in columns 44–47.[56] The following terms occur in ***P.Rev.*** for oil production:

oil factory	ἐλαιούργιον, ἐργαστήριον
workers	ἐλαιουργοί
mortar	ὅλμος
pounders	κοπεῖς
press	ἱπωτήριον, ὄργανον
equipment	κατασκευή

P. Hönigsberg, "Römischen Ölmühlen mahlen noch in Oberägypten," *MDAI* Kairo 18 (1962), pp. 70ff; and D. Bonneau, "Proposition de renouvellement de bail d'une huilerie" *Scritti in onore di Orsolina Montevecchi* (Bologna: CLUEB, 1981), pp. 49–57.

[56] The other references to oil factories in ***P.Rev.*** are: 49.6, 13, 16; 50.20, 23–24; 51.1, 15; 55.11, 20; 56.19; 58.6 (= 60.12).

On the basis of this information from ***P.Rev.*** and on other Ptolemaic papyri, in comparison with evidence from Roman times, the basic workings of the Ptolemaic oil factory can now be determined.

Preparation of Seeds. ***P.Rev.*** 39.8–12 provides the oil seed cultivators an option and a surcharge if they choose the option: if they prefer not to deliver to the oil factories seed that is ready for the mortars (ὅλμος), they may turn in seed as long as it has been cleaned with a sieve (καθαίρειν κοσκίνωι); but for the more thorough cleaning (ἀποκάθαρσις), which will be necessary before it can be crushed, they are to measure from the threshing floor (ἀπὸ τῆς ἅλω) an additional seven artabas per 100 of sesame and castor, and eight of safflower. This passage suggests some essential steps prior to oil production.

Some light threshing of oil seeds is to be expected to release the seeds from the pods (below, ch. 3 §1, p. 57). Thus, the mention of threshing floor in ***P.Rev.*** is consistent with other information. The idea that oil seeds would be threshed was too quickly dismissed by Grenfell;[57] he thought ἅλως would here only be a place of storage, and in support he noted ***P.Rev.*** 41.19 where seed for planting is allotted ἀπὸ τῆς ἅλω, and ***P.Petr.*** II 38a.23 where it is reported that a threshing floor was being used to store hay. It is possible that a threshing floor might become a temporary storage place, but that does not mean that it was not also a place of actual threshing:

1. A threshing floor is distinguished from a granary (θησαυρός) in several papyri.[58]
2. Dioscorides 4.161(164) reports that castor seed is hulled (λεπίζω).
3. ***P.Rev.*** indicates that sesame seed needed to be separated from something (probably chaff) that a sieve would remove.
4. Modern practices, in some cases, include threshing.

Therefore, oil seeds were probably threshed as the term ἅλως suggests. It should be noted that a threshing floor does not necessarily involve animals dragging heavy sledges, however appropriate that may be for some grains. Pliny *Nat.* 18.72(298) describes various means of threshing, including flailing.

Sifting is a logical step after threshing. Galen 13.953 specifies that oil to be used in medicine must have no leaves (θαλλούς) in it when it

[57] Grenfell, *Revenue Laws.* p. 126; cf. Schnebel, *Landwirtschaft*, pp. 171–72.
[58] ***P.Fam.Teb.*** 44.8; 45.6; 47.17; ***P.Tebt.*** III pt. I 727.8; 786.12; 789.12.

is manufactured. In addition to ***P.Rev.***, several papyri suggest that oil seeds were sifted. ***P.Cair.Zen.*** III 59494 is a report of Petearmotis, who calls himself a sifter (κοσκινευτής). He includes sesame and castor in his report of sifting. ***P.Cair.Zen.*** IV 59732 is an account of 1,580 artabas of sesame in three different grades:

καθαρόν	1,311	½ ⅓ art.
ἀπὸ σκυβάλων[59]	250	art.
ἀπόσηστον	18	⅙ art.

The last entry is of sesame that had been sifted.[60] A Roman text (***P.Ryl.*** II 168.12) describes a rent payment in vegetable seed as νέα καθαρὰ ἄδολα κεκοσκινευμένα. Sifting is especially common for grains: a formula for quality control required before the grain could be removed from the threshing floor reads: νέον καθαρὸν ἄδολον ἄκριθον κεκοσκινευμένον.[61]

According to ***P.Rev.*** 39, sifting is the minimum requirement of each cultivator; but a more thorough cleaning (ἀποκάθαρσις) is required before the seed is ready for crushing. Only three other occurrences of ἀποκάθαρσις have been found, each in a context of seeds or grains, and none with any indication of the specific nature of the process.[62] A suggestion for the ἀποκάθαρσις is the use of a water bath. Dioscorides 1.32(38) mentions simply that they clean it (καθαίρω), but Pliny *Nat.* 18.23(98) mentions putting sesame in cold water to separate the chaff from the seed. Also, ***P.Cair.Zen.*** IV 59562.18 mentions ten artabas of sesame seed that had been washed (πεπλυμένον), though this sesame was probably not to be pressed for its oil since it was listed among some foods intended for a festival. Perhaps, then, the ἀποκάθαρσις referred to washing the oil seed.

59 Cf. ***P.Mich.*** I 31.15; and ***P.Cair.Zen.*** III 59494.16.

60 The only known occurrence of this adjective; from the verb ἀποσήθω.

61 ***P.Oxy.*** I 101.37; VI 910.32; 988; VII 1040.14; VIII 1124.11; XVIII 2188.5; XXII 2351.45; ***P.Princ.*** III 147.23; ***PSI*** VII 739.20; ***P.Fuad.Univ.*** 43.42; ***P.Mert.*** II 68.26. See L. A. Moritz, *Grain Mills and Flour in Classical Antiquity* (Oxford: Clarendon, 1958), pp. 159–60.

62 ***P.Cair.Zen.*** II 59292.371; ***P.Lond.*** VII 1995.175; ***Chrest.Wilck.*** II 198.19. Many occurrences of κάθαρσις have been found, especially in the third century B.C.; four occurrences are in reference to sesame: ***P.Hib.*** I 119.19; ***P.Tebt.*** III pt. II 844.3; 1010.3; ***P.Lond.*** VII 1996.43; only two occurrences are later than the third century B.C.; ***P.Ryl.*** II 71 (first century) and ***P.Lond.*** III 1225 (first century A.D.); κάθαρσις is often associated with sifting in the phrase εἰς τὰς καθάρσεις καὶ τὸ κοσκινευτικόν; examples of the occurrence of this phrase: ***P.Petr.*** III 76(v)ii.1; 81.4; 87a(v)i.13, 20; 129a.14; ***P.Lille*** 20.3 *et saepe;* ***P.Ryl.*** II 71.9 *et saepe;* ***P.Tebt.*** I 61b.392; 72.398; 92.10; III pt. II 837.10; 847.15.

In ***P.Cair.Zen.*** IV 59670, a quantity of oil seed that had not been cleaned is called κροτῶνος ῥυπαροῦ. LSJ equates ῥυπαρός here with ἀδειγμάτιστος. With support from a bilingual text, it can be shown that ῥυπαρός should be understood as the opposite of καθαρός.[63] This fits well with the foregoing discussion of cleaning the seed. Perhaps the linking of ῥυπαρός with ἀδειγμάτιστος has some merit, though with a different result; for when the seed was clean, it was verified or stamped. Thus ἀδειγμάτιστος, like ῥυπαρός, indicates oil seed not yet clean. Support for this is in ***Pap.Lugd.Bat.*** 13 where a surcharge of ten artabas per 100 is assessed for δειγματισμόν. The point was not the cost of stamping but of cleaning; as in ***P.Rev.***, the surcharge was for ἀποκάθαρσις.

Production of Oil. Two terms in ***P.Rev.*** are used for the oil factories: ἐργαστήριον and ἐλαιούργιον. The term ἐργαστήριον, meaning oil factory, is exclusive to ***P.Rev.***,[64] while ἐλαιούργιον is found in several Ptolemaic papyri.[65]

The first stage of processing in the oil factory is suggested by ***P.Rev.***'s use of ὅλμος and κοπεῖς. The need for this crushing of the seeds is confirmed by literary sources and by modern processing practices. Further, evidence exists that the seed was rubbed by hand: ***P.Col.*** III 53.3 has λελέανται τῆι χεῖρι.[66] As noted for Roman papyri, a worker is called τριβεύς, and Pliny *Nat.* 18.23(98) says that sesame seed is rubbed (*confrico*).

Evidence for the number of mortars at one factory comes from ***P.Tebt.*** III pt. II 844 (below, ch. 3 §9, pp. 67–68). A factory reports the amount of sesame processed: in one month as much as 188 artabas were manufactured into oil. On the basis of the requirement in ***P.Rev.*** 46.16 that no less than one artaba of sesame be processed per day at each mortar, if this factory met its quota each day at each mortar, it had six mortars.

Secondly, the crushed seeds were pressed. ***P.Rev.*** has two terms: ἰπωτήριον[67] refers to the presses held by temples, while ὄργανον

[63] P. W. Pestman on ***Pap.Lugd.Bat.*** XX 12, n. i.

[64] In the Zenon archive, ἐργαστήριον means brewery five times and pottery shop once: ***P.Cair.Zen.*** II 59199.6; III 59403.6; IV 59742.18; ***P.Mich.*** I 36.10; ***P.Col.*** III 34.7, 9; 52.7. In other Ptolemaic papyri it is most often "grain mill": e.g. ***P.Lille.*** 19.3; 21.20; 23.19; ***P.Tebt.*** I 111.3; 159; 186.

[65] ***P.Cair.Zen.*** II 59223.7; 59247.9; III 59412.6; IV 59717.7; ***P.Hib.*** I 43.7–8; II 245.8; ***P.Lond.*** II 227b.7; ***P.Petr.*** II 27(2).15; ***P.Ryl.*** II 260.3–4; ***PSI*** IV 438; ***P.Tebt.*** III pt. II 844.16 *et saepe;* ***O.Wilck.*** II 737.2; 741.2; 743.2. The locations given for these factories include: Arsinoite, Oxyrhynchite, Aphroditopolite, and Cynopolite nomes, Thebes, and Cynopolis; ***P.Rev.*** 58.6 (= 60.12) mentions a factory at Alexandria. Though ἐλαιούργιον is not extant in ***P.Tebt.*** III pt. I 728, the text clearly refers to a factory.

[66] Cf. Pliny *Nat.* 18.23(98); on λεαίνω see ***P.Berl.Moller*** 13i.4 (= ***SB*** IV 7350.4).

[67] This term occurs only in ***P.Rev.*** among the Ptolemaic papyri.

occurs in other contexts. What difference this suggests, if any, is unknown. The papyrological evidence, at least, suggests that Ptolemaic oil was produced by pressing and not by boiling as Herodotus and Dioscorides report. The only other term for the equipment of Ptolemaic oil factories is κατασκευή, occurring only in ***P.Rev.*** The oil workers are called ἐλαιουργοί in ***P.Rev.***, and in ***P.Tebt.*** I 5.173. In ***UPZ*** I 119.35, they are called κικιουργοί.

Information about Ptolemaic oil factories is not extant in the detail that literary sources or Roman papyri provide. It is sufficient to demonstrate, however, that they were processing the seed in basically the same way. Note the following parallel terms for oil production:

Ptolemaic papyri	*Roman papyri*	*Literary sources*
ἐλαιούργιον	ἐλαιούργιον	
ὅλμος	θυεία	ὅλμον
κοπεῖς		κόπτω
λεαίνω	τριβεύς	*confrico*
ἱπωτήριον		ἀπιπόω, ἀποθλίβω, ἐκθλίβω, *exprimo*
ὄργανον	ὄργανον	ὄργανον
κατασκευή	ἐπισκευή	

§ 5 *Unspecified Oil*[68]

Numerous occurrences of ἔλαιον in the papyri stand without an adjective or an explanatory context. Grenfell, relying on ***P.Rev.***, concluded, "Where ἔλαιον is found in the papyri of this period, meaning one kind of oil, the presumption is that sesame oil is meant."[69] More often than not, editors have followed Grenfell's lead.[70] Some exceptions are Rostovtzeff ("The term ἔλαιον [is used] in the accounts . . . for all the vegetable oils"[71]), Turner (he takes ἔλαιον as olive oil in a third century B.C. papyrus[72]), and Tait

[68] An earlier version of this section, dealing only with the third century B.C., is in *Atti del XVII congresso internazionale di papirologia* (Naples, 1984), pp. 1318–23.

[69] Grenfell, *Revenue Laws*, p. 132.

[70] E.g. Westermann and Kraemer on ***P.Corn.*** 1; Grenfell, Hunt, and Smyly on ***P.Tebt.*** I 116; Reil, *Beiträge*, p. 138.

[71] Rostovtzeff, *SEHHW* I, p. 356.

[72] E. G. Turner on ***P.Hib.*** II 248 (p. 157); I have also previously argued that ἔλαιον without a qualifying adjective meant olive oil: "Oils in Ptolemaic Egypt: The Provisions of ***P.Rev.*** in Light of the Papyri" (Ph.D. diss., Duke University, 1977), pp. 85–87.

("ἔλαιον as is generally recognized can signify either 'olive oil' or 'oil' in general"[73]).

Essential to conclusions about the oils of Ptolemaic Egypt is a determination of the semantic range of ἔλαιον. On the basis of the oils in use, it is likely that ἔλαιον designated one of four possibilities: 1) any oily substance, without specifying the source of that substance; 2) sesame oil; 3) olive oil; 4) safflower oil. Castor oil, though in common use, was always κίκι (below, ch. 2 §6, p. 43) and was carefully distinguished from other vegetable oils.

Literary Sources. Most commonly, ἔλαιον referred to olive oil. Some authors, especially in medical texts, attached adjectives to ἔλαιον that removed from ἔλαιον the specification of olive oil:

Hippocrates	8.306, 308, 386 8.376	ἔλαιον ῥόδινον χηνεῖον ἔλαιον	Va
Aristotle *HA*	520a.18	ἔλαιον ἀπὸ σελαχῶν	IVa
Dioscorides	1.32(38) 1.33(39)	κίκινον ἔλαιον ἀμυγδάλινον ἔλαιον	Ip

In order to differentiate olive oil from other oils, Dioscorides called olive oil ἔλαιον κοινόν. In describing the process of extracting castor oil, he used ἔλαιον for the oil that is skimmed off after boiling. In the Septuagint, ἔλαιον translates either שֶׁמֶן "olive oil" or יִצְהָר "oil." Once, ἔλαιον occurs with the adjective σμύρνινον for oil of myrrh (Esther 2:12). Thus ἔλαιον sometimes designates oils other than olive.[74]

Occurrences in Papyri. The list of Ptolemaic papyri mentioning ἔλαιον (see appendix B) includes numerous accounts, receipts, and letters. Most frequent, especially in the Zenon archive, are lists of various supplies in which ἔλαιον is found without specification.

In ***P.Rev.***, ἔλαιον occurs thirty-four times (see table 1): ἔλαιον can mean sesame oil when an adjective or a context so indicates, it can mean olive oil in the sense of imported oil, and it can refer to more than one oil.

In the Zenon archive, fifty-two documents mention ἔλαιον. In at least twenty-four of those documents there is no indication of which oil is intended. This is true in a few cases because the texts are fragmentary, but in most cases because ἔλαιον was simply given

[73] W. J. Tait on ***Pap.Lugd.Bat.*** XX 25.
[74] Cf. Till, *Arzneikunde*, pp. 80–82.

TABLE 1 *Occurrences of ἔλαιον in Third Century Papyri*

P.Rev.	Use of term	Number of Times
1.	Refers to three or more oils designated in the context	4
2.	Refers to "oil" in general	6
3.	With the adjective σησάμινον	6
4.	Context indicates sesame oil	3
5.	In the phrase ἔλαιον καὶ κίκι	5
6.	In the phrase ἔλαιον σησάμινον καὶ κίκι	1
7.	With the adjective ξενικόν	2
8.	Context indicates foreign oil	5
9.	Distinguished from κίκι in a list	6

Zenon archive

	Use of term	Number of Times
1.	With the adjective σησάμινον	1
2.	In the phrase ἔλαιον καί κίκι	2
3.	Context indicates foreign oil	6
4.	Distinguished from κίκι in a list	5
5.	With the adjective ὀμφάκινος	2
6.	With the adjective λευκός	2
7.	In a list that mentions olives	8

Other papyri

	Use of term	Number of Times
1.	In a context of sesame	1
2.	In the phrase ἔλαιον καὶ κίκι	1
3.	In the phrase ἔλαιον σησάμινον καὶ κίκι	1
4.	Context indicates foreign oil	2
5.	In a list that mentions olives	1
6.	Distinguished from κίκι in a list	6
7.	From a vineyard (ἄμπελος)	1

without specification. For the occurrences of ἔλαιον that may indicate a specific oil, see table 1.

Forty-three other third century documents mention ἔλαιον. While twenty-nine do not suggest the type of oil, thirteen occurrences provide some indication (see table 1).

In the Serapeum archive of the second century, ἔλαιον occurs twelve times with the adjective σησάμινον, nine times in the phrase ἔλαιον σησάμινον καὶ κίκι, and twenty times in the phrase ἔλαιον καὶ κίκι.

Among the forty-three other second and first century texts, one reference to Hellenic oil, one to foreign oil, and two references to "kolpitic" oil are the only designations extant.

Questions about ἔλαιον. Several questions raised by the foregoing tabulations are important for understanding the uses of the term ἔλαιον.

1. Does ἔλαιον in a list in which olives are mentioned indicate that ἔλαιον is olive oil?

Numerous lists record ordinary provisions needed for daily life: oil, meat, fish, olives, etc.[75] Though ἔλαιον occurs in the same list that includes olives, the two terms are usually separated by other necessary commodities. The term ἔλαιον is not likely to specify olive oil in such circumstances, any more than a modern grocery list of meat, flour, and eggs means that the meat to be purchased is chicken. If the authors of these lists had a specific oil in mind, they did not feel it necessary to indicate which.

2. Does ἔλαιον in a list in which κίκι is mentioned indicate that ἔλαιον is castor oil?

Again, these are lists of household provisions; sometimes ἔλαιον occurs in close proximity to κίκι. As indicated above, castor oil is always designated κίκι; and no evidence has been found that would allow ἔλαιον to include castor oil. Apparently κίκι is sufficiently unique to merit specific identification, while ἔλαιον in these lists is unspecified.

3. Does the adjective ὀμφάκινος with ἔλαιον specify olive oil?

This adjective and a related adjective, ὀμφάκιος, occur six times in the papyri:

P.Cair.Zen.	IV	59681	oil	258–57a
		59682	oil	258–57a
P.Flor.		140	wine	IIIp
P.Holm.[76]		281	oil	III–IVp

[75] E.g. ***P.Mich.*** I 2.10; ***PSI*** IV 428.*passim;* V 535.*passim;* VI 553.8–9; 594.10–11; VII 858.17–18.

[76] See Halleux, *Les Alchimistes Grecs*, p. 121.

P.Tebt.	II	273	oil	II–IIIp
P. Turner		14	oil	IIp

In Dioscorides 1.30(29), ὀμφάκινος describes an oil of particular value: ἔλαιον πρὸς τὴν ὑγιείᾳ χρῆσιν ἄριστον τὸ ὠμοτριβές, ὃ καὶ ὀμφάκινον καλοῦσι.[77] And Galen 13.953 prescribes a concoction, with one ingredient being oil that could not be ὀμφάκινος nor that had leaves (θαλλούς) mixed in with the fruit when it was processed. Of these references, ***P.Turner*** 14 provides the clearest evidence: τῆς ἐλαίας ἀποθλιβόμενον ἔλαιον τὴν ἐναντίαν κεχώρηκεν. ποιεῖται μὲν γὰρ ὀμφάκινον σταλτικόν. Hence ὀμφάκινος in a context of oil designates unripe olive oil.

4. Does the adjective λευκός with ἔλαιον specify olive oil?

Dioscorides 1.30(32) describes the process for making ἔλαιον λευκόν: it involves bleaching in the sun, repeatedly pouring from container to container, stirring every day, and adding various ingredients. Since Dioscorides's arrangement is topical, he apparently intends this process to apply to the oil under consideration, olive oil, though he does not explicitly limit its application to one oil. Hippocrates 8.308 prescribes a cure using αἰγύπτιον ἔλαιον λευκόν. The two papyri that mention ἔλαιον λευκόν are, however, about imported oil (***P.Cair.Zen.*** 59012, 59013). The adjective λευκός suggests, then, a specially refined form of oil, probably olive oil.

5. Does the phrase ἔλαιον καὶ κίκι provide identification of ἔλαιον?

The phrase ἔλαιον καὶ κίκι occurs five times in ***P.Rev.*** and two times in the other third century papyri, while ἔλαιον σησάμινον καὶ κίκι occurs twice. Column 41 of ***P.Rev.*** provides the following regulation: in the event the required amount of sesame and castor seed is not planted, the penalty is as follows: 2 dr. are to be paid to the tax farmer per each artaba of sesame and castor seed and whatever the profit would have been τοῦ ἐλαίου καὶ τοῦ κίκιος. In column 49, a corrector added καὶ κνῆκον to ἔλαιον καὶ κίκι and inserted σησάμινον after ἔλαιον; this may suggest that ἔλαιον was not ambiguous with καὶ κίκι, but by extending the string to καὶ κνῆκον it became unclear. Also in ***P.Rev.***, column 51, the temples are allowed

77 Hippocrates, Pliny, and Galen also use ὀμφάκινος for unripe olive oil; see LSJ for references. Dioscorides 1.39–40, 44, 46, 48–49, 62 gives it as an ingredient of various medicines.

to acquire sesame and castor oil, but a record τοῦ κίκιος καὶ τοῦ ἐλαίου must be kept. So Grenfell concluded: "ἔλαιον, when coupled with κίκι, means sesame oil. . . ."[78]

Additional evidence comes from the second century Serapeum collection which has repeated references to amounts of sesame and castor oil for the temple. Twenty-four papyri are petitions by Ptolemaios, son of Glaucias, on behalf of twin girls living in the Serapeum, for one metretes of sesame oil and one of castor oil per year that was due the twins. These allowances were withheld in the years 164–62 for unknown reasons, and Ptolemaios attempted to have the allowances reinstated. The two oils are referred to in various ways, sometimes separately, ἔλαιον σησάμινον and κίκι, but usually together, ἔλαιον σησάμινον καὶ κίκι, or most commonly, ἔλαιον καὶ κίκι. In ***P.Rev.*** and in the Serapeum archive, therefore, ἔλαιον is used in the phrase ἔλαιον καὶ κίκι to designate sesame.[79]

Cognates and Synonyms of ἔλαιον. Several words related to ἔλαιον are generally used for oil-related matters. The adjective ἐλαϊκός is a common term in ***P.Rev.*** and in other papyri for contracts and other things pertaining to more than one of the oils.[80] Likewise, the oil factories, the oil workers (ἐλαιουργία, ἐλαιουργός), and the oil dealers (ἐλαιοκάπηλος) handle a variety of oil sources, as three papyri,[81] reporting a delivery of both sesame and castor seed to an oil factory, indicate.

Words that can apparently also be used without specific reference to a particular oil are ἄλειμμα and ἐλλύχνιον (or ἐπελλύχνιον). A gymnasiarch is praised for supplying ἄλειμμα in ***SB*** 7246.19 (III–II B.C.). The occurrences of ἐλλύχνιον (or ἐπελλύχνιον) are:

P.Rev.		40.10, 12	259a
		55.9	
P.Mil.	II	27ii.26, 31	IIa
P.Tebt.	III pt. II	894(7).5	IIa
UPZ	I	101.6	IIa
	II	204.3–5	IIa
P.Lond.	I	121.376	IIIp
Stud.Pal	*XXII*	56.17	II–III

78 In the introduction to ***P.Cair.Zen.*** II 59187 (255), Edgar takes ἔλαιον καὶ κίκι as "table oil and lamp oil." Since κίκι was probably not used in cooking and since ἔλαιον could be a general term for oil, that suggestion is attractive. The evidence above, however, argues against Edgar's proposal.

79 Grenfell, *Revenue Laws*, p. 131.

80 E.g. ***BGU*** VI 1219.25,38: ἐλαϊκῶν φορτίων.

81 ***P.Cair.Zen.*** III 59412; ***P.Tebt.*** III pt. I 844; ***P.Hib.*** II 245.

The three uses in ***P.Rev.***, however, seem to refer specifically to linseed oil, for ἐπελλύχνιον replaces ἔλαιον τὸ ἀπὸ τοῦ λίνου σπέρματος, perhaps to save space.

Conclusion. A solution for the designation of ἔλαιον is based on broad evidence:

1. ἔλαιον is used as a general term for oil in Greek literary texts.
2. ἔλαιον is used as a general term for oil in ***P.Rev.***
3. Olive oil or sesame oil is at times specified by appropriate adjectives.
4. ἔλαιον is carefully distinguished from κίκι but not from other types of oil.
5. The cognates of ἔλαιον apply to more than one oil.
6. Lists that mention ἔλαιον often have other general terms: e.g. κρέας (meat), ὄψον (fish), σῖτος (grain).
7. Mutually exclusive adjectives sometimes modify ἔλαιον, suggesting that ἔλαιον has a range of meaning broader than one oil.

The uses of ἔλαιον, therefore, are three:

1. It is a general term for oil with a semantic range broad enough to encompass any or all of the vegetable oils except castor.
2. It designates a specific oil when an adjective indicates that a particular oil is intended.
3. It designates sesame oil in the phrase ἔλαιον καὶ κίκι.

§ 6 *Imported Oil*

According to columns 52–54 of ***P.Rev.***, the government put strict limits on oil that was imported into Egypt. It banned the sale of foreign oil anywhere in Egypt except at Alexandria and at the port, Pelusium. An individual was permitted, however, to transport imported oil anywhere in Egypt; but only for his personal use, and only if he registered it upon his transporting it out of Alexandria or Pelusium, obtained a voucher, and paid a tax of 12 dr. per metr. (the taxes collected were to be forwarded to the nome to which the oil was destined). Businessmen were permitted to transport oil from Pelusium to Alexandria without paying the tax, but they were required to have a voucher. As a precaution, tax contractors were instructed to have agents posted at Pelusium and Alexandria, to have them record all amounts of oil shipped in, and to supervise the storage of the oil until it was issued. Penalties are stated at 100 dr. per metr. plus the confiscation of the oil for any individual transporting unauthorized oil; and for the businessman who transported oil without a voucher, the penalty was the confiscation of his oil.

Terminology. Two terms in ***P.Rev.*** designate imported oil: ξενικόν and Σύρον.[82] Though the terms may differentiate between oil from non-Ptolemaic dominions and oil from a Ptolemaic dominion, the regulations are the same for both. Other Ptolemaic papyri have the terms:

ξενικόν	***P.Cair.Zen.***	I	59012.12, 93, 104, 114	259
			59013.13	259
			59015	259–58
			59077.2, 9	257
	P.Lond.	VII	2162.6	III
	P.Tebt.	III pt. I	728.8	II
		pt. II	887.*passim*	early II
			997.4–5	early II
	UPZ	II	228r.2, 5; v.2	135–34
	P.Tebt.	I	121.100	94/61
			253	96/63
Σύρον	***P.Hib.***	II	248iii.3	c.250
	P.Petr.	III	47a.4; b.4	III
Κρητικόν	***P.Tebt.***	III pt. II	997.4–5	early II
Ἑλληνικόν	***O.Bodl.***	I	320v.10	I

P.Cair.Zen. 59015 reports a shipment of foreign oil in Milesian (cf. ***P.Tebt.*** II 248iii.3) and Samian jars. ***P.Cair.Zen.*** IV 59544.2(257a) is a letter to Apollonius reporting the shipment of two petalia of olives. In the margin is Ἑ̣λ̣λ̣ηνι̣κ̣[ῶν] about which the editor says "letters mutilated but certain." These references appear to be clear indications of imported oil; they are not just brands of oil. Syrian and Cretan oil occur in connection with foreign oil, and the Maroetic harbor at Alexandria and the harbor at Pelusium are mentioned as the ports where the oil arrived.

The adjective κολπιτεικόν appears in several papyri from Kerkeosiris and has been thought to indicate oil imported from Syria (***P.Tebt.*** I 38.12; 125; IV 1094.3). Grenfell, Hunt, and Smyly proposed

[82] Ξενικόν: 52.10, 13, 22, 24; 53.2; 54.16. Σύρον: 52.26; 54.17. Rostovtzeff tried to see a distinction between "foreign" and "Syrian" and the customs assessed oil imported from Ptolemaic dominions and those assessed the oil imported from non-Ptolemaic dominions. I do not think a case can be made for this from ***P.Rev.*** *SEHHW* III, p. 1417, and "Foreign Commerce of Ptolemaic Egypt," *Journal of Economic and Business History* 4:4 (1932) 764–68. On Σύρον, see V. Tcherikover, "Palestine under the Ptolemies (A Contribution to the Study of the Zenon Papyri)," *Mizraim: Journal*

that kolpitic meant oil from Syria on the grounds that ***P.Rev.*** cites Syria as "the chief source of oil imported to Egypt."[83] Additional support is in Stephanus of Byzantium who lists kolpitic under Φοινίκη. Nothing in these three texts, however, requires this to be imported oil. Furthermore, the verbs κολπιτεύω and διακολπιτεύω in ***P.Phil.*** 35.22 and in ***P.Tebt.*** 709.9 mean smuggle.[84] Thus kolpitic may mean only contraband,[85] and not Syrian oil. Since ἔλαιον καὶ κίκι designates sesame and castor oil (above, §5, pp. 22–23), then ***P.Tebt.*** I 38.12 refers to contraband sesame and castor oil (κολ[πιτ]ικὸν ἔλαιον καὶ κίκι), certainly not to Syrian sesame and castor oil.

P.Cair.Zen. IV 59583 and ***P.Col.Zen.*** I 14 were joined by Edgar and published with the conclusion that the letter concerned the arrival of a shipment of imported oil at Alexandria.[86] The decision was based on ἐξ οἴκου which Edgar took to mean "from our own property (overseas)." Tait has shown, however, that οἶκος more properly refers to Apollonius's Egyptian estates and thus not to foreign oil.[87]

Amounts, Prices, and Duties. In addition to ***P.Rev.***, several papyri are instructive on the enterprise of importing oil:

P.Cair.Zen. I 59012 259

The retail value, import duties, and taxes are given for various imported goods arriving at Pelusium; two entries list a hemikadion of ἔλαιον λευκόν valued at 30 dr., charged a 50% import duty, and assessed two additional taxes; one entry lists a φορω[88] of ἔλαιον valued at 25 dr., charged a 50% import duty, and assessed taxes of 1 ob. and 1 dr., 2 ob.; and one entry lists two φορω of ἔλαιον valued at 50 dr., charged a 50% import duty, and assessed taxes of 2 ob. and 2 dr., 4 ob.

---

of Papyrology, Egyptology, History of Ancient Laws, and Their Relations to the Civilizations of Bible Lands 4–5 (1937) 14–15.

[83] See the footnote on ***P.Tebt.*** I 38.12; so also H. Maspero, *Les finances de l'Egypte sous les Lagides* (Paris, 1905), p. 74.

[84] M. Rostovtzeff, "Review of N. Lewis, *L'industrie du papyrus dans l'Egypte Grecoromaine,*" *Gnomon* 12 (1936) 51–52; see also LSJ *Supplement* (1968), s.v.; J. G. Keenan and J. C. Shelton on ***P.Tebt.*** IV 1094.3; and Rostovtzeff, *SEHHW* I, p. 355.

[85] Two other papyri known to mention illegal oil are: ***P.Tebt.*** I 39.9–10 (ἐλαϊκὸν ἐπίτιμον) and ***P.Hib.*** I 59.7 (τὸ κ̣λ̣ἐ̣πιμον ἔλαιον).

[86] C. C. Edgar, "A New Letter of Apollonios the Dioketes," *Archiv für Papyrusforschung* 11 (1935) 218–19; for republication, see following footnote.

[87] ***Pap.Lugd.Bat.*** XX 25.

[88] This occurs only in the Zenon archive (three times) and is a Demotic measure: Pestman, ***Pap.Lugd.Bat.*** XX 12.7, 21.

P.Cair.Zen. I 59013 259
Shipping charges are given for goods being transported from the Maroetic harbor to Alexandria; for 1 hemikadion of ἔλαιον λευκόν the charge was 1 ob.

P.Cair.Zen. I 59015 259–58
Calculations are given regarding import of a large quantity of oil: total amount was nominally 1,000 metr. shipped in Milesian and Samian jars, 526 keramia and 278 hemikadia; this was equivalent to 665 keramia, which contained on the average 16 ch. (not 18), hence the total amount was more accurately 888 8/9 metr. But due to some breakage, the amount received was 881 metr., 4 ch. At a market price of 52 dr. per metr., the oil was worth 7 talents, 3,812 dr.; but an import duty of 50% plus freight charges and taxes amounted to 4 talents, 637 dr. The oil was delivered to the government storerooms at Alexandria, for which the government paid 46 dr. per metr.[89]

P.Lond. VII 2162 III
An import charge of 18 dr. is listed for 2 hemikadia of oil.

P.Hib. II 248 iii c.250
Letter about 3 metr. of Syrian oil; Oxyrhynchite nome.

P.Petr. III 47 III
Two receipts for 2 1/2 measures of Syrian oil, valued at 5 dr. each for a total of 12 dr. 3 ob., supplied to stone cutters.

P.Tebt. III pt. I 728 II
A deficiency is reported to a government official: 47 metr., 3 ch., 8 kotylai of foreign oil.

P.Tebt. III pt. II 887 early II
Receipts by an oil merchant of small quantities of foreign oil usually valued at 80 dr. per kotyle (=5,760 dr. per 6 ch. metr.). His sales in an eight day period totaled more than 63 kotylai.

[89] For full discussion of the implications of all this see C. C. Edgar, "Selected Papyri from the Archives of Zenon," *Annales du service des antiquités de l'Égypte* XXIII (1923) 86–95.

P.Tebt.	III pt. II	997	early II

Receipts of 2 metr. of foreign oil and 1 metr. of Cretan oil.

UPZ	II	228	135–34

A receipt of 80 metr. of foreign oil with a transportation fee of 10 dr. per metr. for a total of 800 dr.; Thebes.

P.Tebt.	I	121	94/61

An expenditure of 400 dr. for foreign oil.

P.Tebt.	I	253	96/63

A record of 100 ? of foreign oil.

Some fluctuation in the value of imported oil is obvious from the figures in the above papyri (the equivalency is given in a 6 ch. metr.):[90]

30 dr./hemikadion (60 dr./metr.)
52 dr./metr.
18 dr./hemikadion (36 dr./metr.)
80 dr./kotyle (5,760 dr./metr.)

The first three of these pricings are from the Zenon archive, while the last is from the second century and reflects the copper inflation.[91] Converting the 48 dr. per 12 ch. metr. of native oil (approximately 1⅓ larger than a 6 ch. metr.), the oils produced in Egypt are worth 36 dr. (according to ***P.Rev.***), while the imported oil ranged from 36–60 dr. in the third century. These figures reflect an apparent retail market price, which varied according to the quality or type of oil imported; the government only paid 46 dr. per metr. for oil valued at 52 dr.

Other costs regarding imported oil show less variation, perhaps because our sample of evidence is smaller. Import duty was 50% while shipping charges were 1 ob. per hemikadion (third century) and 10 dr. per metr. (second century). The first shipping charge was

[90] A 6 ch. metr. was standard in the Zenon archive; on the various measurements mentioned here, see above §3, pp. 9–10.

[91] On inflation see A. Gara, *Prosdiagraphomena e circolazione monetaria,* Testi e documenti per lo studio dell 'antichita' LXI (Milan: Cisalpino-Goliardica, 1976), pp. 147–57; T. Reekmans, "The Ptolemaic Copper Inflation," *Studia Hellenistica* 7: *Ptolemaica* (1951), pp. 61–118.

for a short distance; the distance of the second is unknown. The taxes assessed imported oil in addition to customs duties totaled approximately 6% of the valuation: using the 25 or 50 dr. quantity in ***P.Cair.Zen.*** I 59012, the first tax was 0.66% and the second 5.33%.[92]

Combining all that an importer of oil might be charged in comparison to what he was paid by the government, if the oil was valued at 52 dr., he paid 26 dr. customs duty, 3 dr. additional taxes, and transportation charges. His costs then, once the oil was in port, were approximately 30 dr. per metr. Since he would be paid 46 dr. by the government, only 16 dr. remain to cover purchase price, shipping from point of origin to Egypt, and profit.[93] Segrè[94] finds twenty-seven figures extant for oil prices in the Greek world, preserved primarily in inscriptions. The figures date from late fourth to early second centuries and are based on a 12 ch. metr. The highest price is 55 dr. per metr., and the lowest 15; an average of the twenty-seven figures is 23.5 dr. per metr. Reducing these figures to a 6 ch. metr., the range is 42.3 down to 11.5 and an average of 18. Oil importers could not then have made a profit if they bought oil abroad at the above retail prices. Other factors must therefore be involved. The imported oil mentioned in the Zenon archive may have come from private estates of Apollonius in Syria, etc. By controlling production, shipping, and importing oil, Apollonius may have made oil importing a successful enterprise.

Need for oil. The motivation for importing oil into Egypt was probably not to satisfy the oil needs of the native population.[95] Castor and other oils were available in adequate supply. But these oils were less desirable to the Greek population. Thus, as ***P.Rev.*** shows, the Ptolemaic government catered to the residents of Alexandria by allowing foreign oil to be marketed there. Consistent with the tenor of ***P.Rev.***, every precaution was taken to protect the success of the oil business in the interior of Egypt. In no way was imported oil to diminish the prospects of profit and taxes from the oil industry.

[92] The smaller of these taxes was for the upkeep of the navy; the larger was an inland toll. See the discussion in A. Andréadès, "Des droits de douane prélevés par les Lagides sur le commerce extérieur," *Mélanges Gustave Glotz* (Paris: Les Presses universitaires, 1932), I, p. 12.

[93] See the discussion in Préaux, *L'économie*, pp. 83–87, and M. Rostovtzeff, "Foreign Commerce of Ptolemaic Egypt," *Journal of Economic and Business History* 4:4 (1932) 764–68.

[94] A. Segrè, *Circolazione monetaria e prezzi nel mondo antico ed in particolare in Egitto* (Roma: Libreria di cultura, 1922), pp. 166–67.

[95] Rostovtzeff, *SEHHW* I, pp. 355–56. Cf. M. I. Finley, *The Ancient Economy* (Berkeley: University of California, 1973), pp. 132–33.

Contrary to ***P.Rev.***'s restriction on foreign oil being sold in the interior of Egypt, the papyri demonstrate that foreign oil was showing up at Thebes, Tebtunis, and the Oxyrhynchite nome; and since receipts for foreign oil were recorded, it is apparent that this oil was not being brought into Egypt by an individual for his own use.

Except for ***P.Cair.Zen.*** I 59015, the amounts of oil imported, however, were small. Imported oil sold at a market price significantly higher than native oil. Even so, the profitability of importing oil was not attractive, given the price of oil abroad and the import duties assessed by the government. Imported oil was therefore a restricted commodity in Ptolemaic Egypt.

§ 7 *Demotic Papyri*

Four terms have been found in the Demotic papyri that can be identified with oils: *nḥḥ* "oil," *ꜣḳj* "sesame," *tgm* "castor," and *gꜣgꜣ* "castor." The precise identification of these terms, however, must be examined.

Oil. The texts that mention *nḥḥ*[96] include a magical text (***P.London-Leiden***), a group of marriage contracts (***P.dem.Eheverträge***, ***P.dem.Ryl.***, ***P.Tor.Amenothes***), two medical papyri (***P.Tebt.Tait.***, ***P.Vindob.D. 6257***), and a variety of receipts and accounts.

The oil *nḥḥ* was common since Pharaonic Egypt and is attested in Hieroglyphic and Hieratic texts as well as Demotic.[97] In his *Demotisches Glossar*, Erichsen translates it "oil" and is followed by some editors.[98] On the other hand, it is also assumed to be olive oil.[99] The most common notion of the last sixty years makes it sesame oil.[100]

The identification of *nḥḥ* with sesame began with Keimer.[101] He assigns the beginning of sesame cultivation in Egypt to the New Kingdom (XIXth dynasty) and concludes that sesame was introduced to Egypt from Asia at the time of the Asiatic conquest.[102]

[96] See appendix C.

[97] Hayes, "Inscriptions," p. 93; Lucas, *Egyptian Materials*, pp. 331–32. For a list of *nḥḥ* in Hieratic texts, see Janssen, *Commodity Prices*, p. 331.

[98] E.g. Tait on ***P.Tebt.Tait*** 18(i).7; Lichtheim on ***P.dem.Medinet Habu*** 148.1; and Nur El-Din in ***P.dem.Leiden.***

[99] Pestman on ***P.Tor.Amenothes*** 1.4; Griffith on ***P.dem.Ryl***. 10.2.

[100] Keimer, *Gartenpflanzen* pp. 19–20; Lüddeckens on ***P.dem.Köln*** 2:20–21; and in ***P.dem.Eheverträge***, p. 262; Janssen, *Commodity Prices*, pp. 330–33; Porten, *Archives*, p. 92.

[101] Keimer, *Gartenpflanzen*, pp. 19–20.

[102] Even if Keimer's general thesis were correct, his date is too late; jars from Tell el Amarna inscribed with *nḥḥ* would suggest an earlier date for the introduction of *nḥḥ;* Hayes, "Inscriptions," p. 93, n. 142.

The evidence reveals, however, that *nḥḥ* is a general term for oil and not a reference to a specific oil:

1. Mutually exclusive adjectives sometimes modify *nḥḥ*, suggesting that *nḥḥ* is a general term. ***P.London-Leiden*** illustrates this: in some cases the quality of oil is specified; in other cases the nature of the oil: moringa, *tesheps,* henna, and vegetable.
2. The evidence for the cultivation of sesame is only found after the arrival of the Greeks.[103]
3. The term *nḥḥ* is only used for oil and never in a context of an oil seed or oil crop, in contrast to terms like *tgm* which can mean seed, crop, or oil.

The strongest argument linking *nḥḥ* with sesame, which was overlooked by Keimer, appears to be the close mention of *nḥḥ* and *tgm*. As suggested above (§5, pp. 22–23), ἔλαιον in the phrase ἔλαιον καὶ κίκι may specify sesame oil. In the Demotic texts, however, though *nḥḥ* and *tgm* occur in the same context in lists of provisions, they are not in a technical phrase like ἔλαιον καὶ κίκι. Thus *nḥḥ,* like ἔλαιον, has a semantic range similar to "oil" in English and cannot be identified with any specific oil except by context or adjective. It was a term that could refer to any of the oils available (e.g. ben or balanos oil in Pharaonic Egypt) except castor.

Sesame. The Demotic occurrences of *ꜣḳj,*[104] firmly identified with sesame,[105] are, with one exception, late Ptolemaic or early Roman. They neither mention large quantities of sesame seed nor show evidence of sesame oil. All but one of these documents (***P.dem.Medinet Habu*** 93) are associated with temples, which are often concerned with oil.[106] ***P.dem.Loeb*** 17, a very early Ptolemaic example of the presence of sesame and the earliest known reference to sesame in Egypt, antedates by half a century the earliest Greek papyrus to mention sesame (***P.Tebt.*** III 845 [264a]).[107]

[103] Alph. de Candolle, *Origine des plantes cultivées* (Paris: Felix Alcan, 1896), p. 338; V. Loret, *La flore Pharaonique* (Paris: Ernest Leroux, 1892), p. 57; G. Schweinfurth in a "mündliche mitteilung" to Keimer (*Gartenpflanzen*, p. 19) says that the cultivation of sesame was first introduced into Egypt in Ptolemaic times; Erichsen, *Demotisches Glossar*, p. 224; Dawson, "Studies," p. 65.

[104] See appendix C.

[105] Erichsen, *Demotisches Glossar*, p. 12; see also the editors of the texts that mention *ꜣḳj*.

[106] E.g. ***UPZ*** I 19–46; above §5, p. 23.

[107] See ch.3 §4, pp. 61–62, for sesame in Mesopotamia and its introduction into Egypt.

Castor. The presence of castor in Egypt is well documented in drawings, in archaeological remains, and in documentary and literary evidence.[108] The Demotic *tgm* has been identified with castor for nearly a century,[109] and with wide acceptance,[110] though a few prefer to translate it as "oil."[111] The identification is certainly correct, for bilingual texts match *tgm* with Greek κροτών.[112]

The texts that mention *tgm*[113] include, in addition to common receipts and accounts, a group of marriage contracts (***P.dem.Eheverträge***, ***P.Tor.Amenothes***) and texts from a temple (***P.dem.Tempeleide***).

Typical of marriage contracts is a list of obligations, including the husband's duty to supply to his wife monthly quotas of oil, occasionally *nḥḥ,* occasionally *tgm,* but usually both.[114] The amounts are 12, 24, or 36 *hin* per year. This suggests two things: first, that oil was important in native culture as one of three common marital obligations—money, wheat or barley, and oil. Oil was considered essential for proper existence. Second, if *nḥḥ* is a general term for oil and if *tgm* specifies castor oil, then the Greek papyri and Demotic papyri are alike in differentiating between oil and castor oil in a list. As suggested above (§5, pp. 21–23), ἔλαιον in a list with κίκι does not specify what ἔλαιον is, as it does in ἔλαιον καὶ κίκι, but only suggests that κίκι is so unique as not to fit under the heading ἔλαιον. Whether this notion was present in Egypt before the Greeks arrived is unclear from the dating of the papyri, for all these Demotic papyri mentioning *tgm* and *nḥḥ* are 230 B.C. or later. Since the Demotic papyri reflect the native Egyptian thinking and since the Greek immigrants had little influence over that,[115] it is safe to conclude that the distinction between *nḥḥ* and *tgm* was a native idea.

Very few occurrences of *gʾgʾ* have been found.[116] Apparently, *gʾgʾ* can be identified with the older term *kꜣkꜣ*, a Hieroglyphic term,

[108] Below, ch. 2 §4, pp. 41–42.

[109] Loret, *Flore*, p. 49.

[110] Grapow/Deines, *Wörterbuch*, p. 584; Spiegelberg on ***P.dem.Zen.*** 1, 3; Griffith on ***P.dem.Ryl.*** 9; Pestman on ***P.Tor.Amenothes*** 1; Wångstedt on ***O.dem.Zürich*** 18–20; Pestman on ***Pap.Lugd.Bat.*** XX 12–13; Dawson, "Studies," pp. 62–68; Porten, *Archives*, pp. 92–93.

[111] Erichsen, *Demotisches Glossar*, p. 662; Thompson on ***O.dem.Theb.*** 6; Kaplony-Heckel on ***P.dem.Tempeleide*** 42, 78–79, 88, 97, 142; Lüddeckens on ***P.dem.Eheverträge*** 10, 19–21, 25, 27, 35.

[112] ***Pap.Lugd.Bat.*** XX 12–13; cf. Keimer, "Naturgeschichte," pp. 100–102.

[113] See appendix C.

[114] Papyri that specify only *nḥḥ:* ***P.dem.Eheverträge*** 13–14, 29; only *tgm:* ***P.dem.Eheverträge*** 10; both: ***P.dem.Eheverträge*** 17, 19–21, 25, 27, 35; ***P.Tor.Amenothes*** 1.4.

[115] H. C. Youtie, "Response" *Scriptiunculae Posteriores* I (Bonn: Rudolf Habelt, 1981), pp. 57–58.

[116] ***I.dem.Nubie*** Ph.417.6: pure oil; ***O.dem.Ossirinco*** 5.2: 4 of oil.

and source of the Greek κίκι as Herodotus (2.94) and other Greek sources suggest.[117] Two Hieroglyphic texts confirm the identification of *k3k3* as castor:

1. A magical papyrus at Turin: "By death owing to trees, by death owing to *k3k3*, by death owing to every kind of reed."
2. "Admonitions of an Egyptian Sage" in the Leiden Papyrus: "The land is as *k3k3* that destroys men."[118]

The reference to the lethal nature of *k3k3* suggests that *k3k3* is castor, for castor seeds are known to be very poisonous.

Summary. The infrequent references to sesame in Demotic papyri mention only small amounts of sesame seed, not more than 3⅚ 1/24 art., and mostly for use in temples. The amounts of castor seed for temples are also small, but outside the temples large amounts occur, up to 100 art. The only indication of pricing is ***P.dem.Zen.*** 1, where it is 1½ *kite* (= 3 dr.) per art.

Quantities of *nḥḥ* vary from ½ *hin* to the common 12 or 24 *hin* in marriage contracts, to 35 *hin;* the latter is priced at 102 silver (pieces) in ***O.dem.Leiden*** 96i.4. The largest quantity of oil noted is castor oil and is also the earliest text: ***P.dem.Ryl.*** 9 is a petition of the sixth century that offers a yearly stipend including 200 *hin* of castor oil. No evidence of the price for castor oil has been found. Sesame oil does not occur, nor is olive or safflower oil mentioned.

[117] Below, ch. 2 §3, pp. 39–40.

[118] Dawson, "Studies," pp. 57–68, cites these papyri but argues that *k3k3* means weed or brushwood, ignoring the lethal nature of castor. Against Dawson are the judgments of Erman, Keimer, and Grapow/Deines, though in some cases the decisions are tentative. The last makes the pertinent observation that *k3k3* as a drug must be a specific plant and not something so general as a bush; A. Erman and H. Grapow, *Wörterbuch der ägyptischen Sprache* (Leipzig: Hinrichs, 1926) 5.109.2, and *Ägyptisches Handwörterbuch* (Berlin: Reuther & Reichard, 1921), p. 194; Keimer, *Gartenpflanzen,* p. 70, 164; Grapow/Deines, *Wörterbuch,* p. 527. Those who are inclined to agree with Dawson are Gardiner, Faulkner, and Burton; A. Gardiner, *The Admonitions of an Egyptian Sage from a Hieratic Papyrus in Leiden* (Hildesheim: Georg Olms, 1969), p. 86; R. O. Faulkner, *A Concise Dictionary of Middle Egyptian* (Oxford: Oxford University, 1962), p. 284; A. Burton, *Diodorus Siculus, Book I, A Commentary,* Études préliminaries aux religions Orientales dans l'empire Romain, Vol. 29 (Leiden: Brill, 1972), p. 133. In regard to the Demotic *gꜣgꜣ*, Erichsen and Charpentier give only "oil," and Bresciani says "*gꜣgꜣ* e un tipo di olio, che solo con dubbio puo essere avvicinato al gr. κίκι"; Erichsen, *Demotisches Glossar,* p. 574; Bresciani on ***O.dem.Ossirinco*** 5, and G. Charpentier, *Recueil de matériaux épigraphiques relatifs a la botanique de l'Égypte antique* (Paris: Trismégiste, 1981) nr. 1288. For a full discussion and refutation of Dawson's arguments, see D. B. Sandy, "Egyptian Terms for Castor," *Chronique d'Égypte* 62: 123–124 (1987) 49–52.

The chronological distribution of the Demotic papyri that mention oils suggests that castor oil was used primarily in the Ptolemaic period and little thereafter. On the other hand, the evenly distributed occurrences of *nḥḥ* reveal that one or more other oils often met the needs of the native population.

II

Castor
(κίκι, κροτών)

§ 1 *Physical Description*[1]

Ricinus communis L. (Fam. Euphorbiaceae) is a branching plant with broad leaves and spike clusters of fruit. In temperate latitudes it is an annual plant as small as 1.0 m., but in the tropical and subtropical areas it is a perennial and grows wild; it often becomes a small tree up to 12 m. in height that lives for many years. The bushy plant has an erect, hollow stem with numerous branches starting just above the surface of the soil. The leaves are very large, dark green, glossy, and have palm-like lobes, such that from the fifteenth to eighteenth centuries, it was usually called "Christ's Palm" or "Palma Christi."[2]

The outer hull of the castor seed[3] is green with soft spines, but as the season progresses, the hull turns brown, splits, and exposes three seeds, each enclosed in a shell. Inside the shells are oval and slightly laterally flattened seeds. Castor seeds vary in length from 8–20 mm., usually weigh 0.3–0.5 gr., and their skin may be red, white,

[1] The most helpful sources on this topic are: Weiss, *Castor, Sesame, and Safflower,* pp. 45–90, 108–30, and Janson, *Castor Oil Production*, pp. 1–13, 19–23; other sources are: Vaughan, *Oil Seeds,* pp. 75–77; Eckey, *Fats and Oils,* pp. 588–91; Usher, *Dictionary of Plants,* p. 507; Trease and Evans, *Pharmacognosy,* pp. 330-33; and Kirschenbauer, *Fats and Oils,* p. 166. For photographs and drawings of castor seed and plant, see Vaughan, and *EB*[15].

[2] P. Canvane, "Dissertation on the Oleum Palmae Christi sive Oleum Ricini (Commonly called Castor Oil) . . ." (2nd ed.; London: Robinson & Roberts, 1769). *EB*[11] 5.482 reports leaves as big as two feet in diameter. Cf. other names for castor in Bedevian, *Plant Names,* p. 511.

[3] At the suggestion of Eckey (*Fats and Oils,* pp. 588–91), the common practice of referring to castor seeds as beans will not be followed: " . . . since the plant is not a legume and the seeds are not edible, in fact violently poisonous."

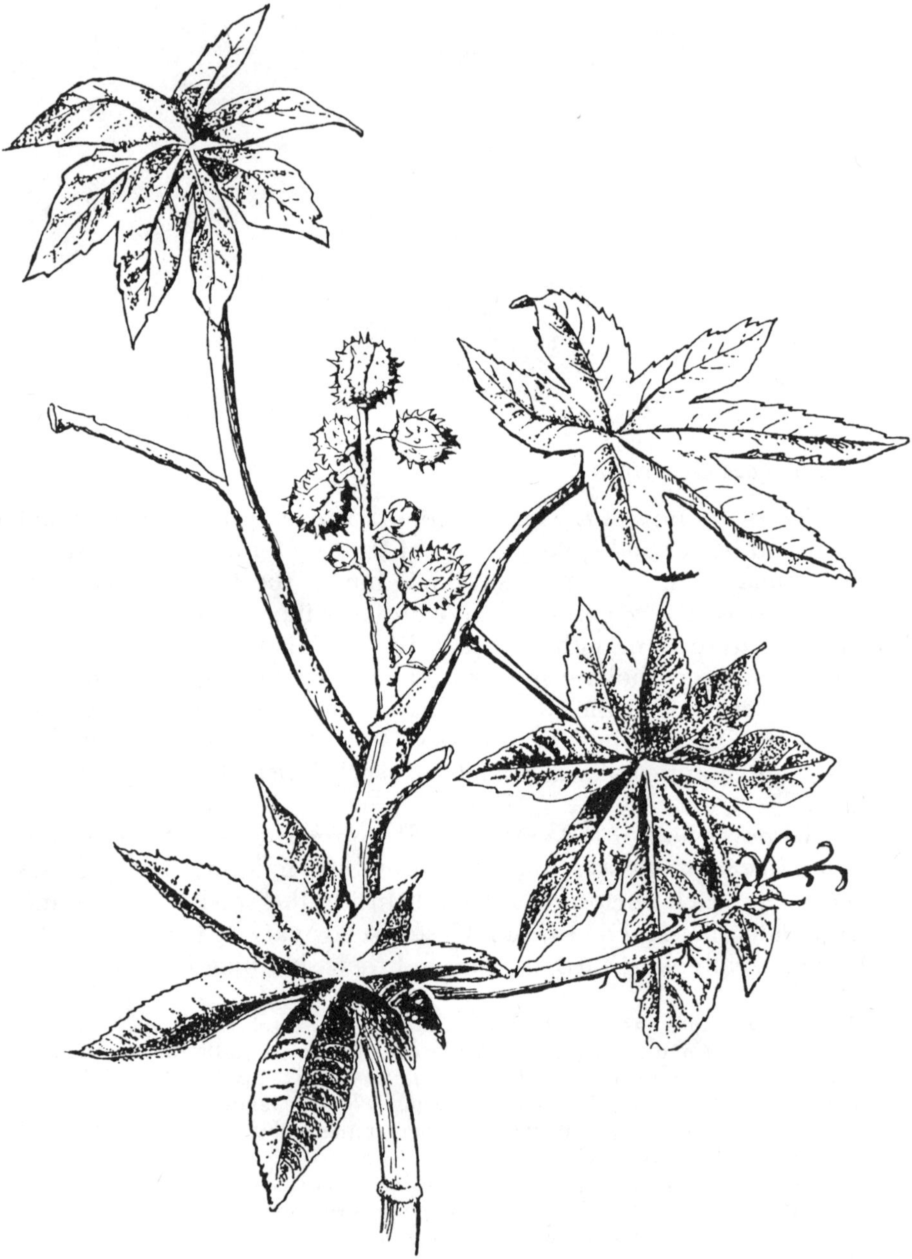

CASTOR

brown, or purple in color. The white inner flesh of the seed is very soft. Castor seeds have a strong smell and are very toxic, but when the oil is extracted, the poisonous properties remain in the meal and the oil is harmless. Castor seed contains 35–55% oil, usually about 50%.[4]

Castor grows well even in unfavorable conditions. It needs water only early in the growing season; afterwards it will be productive without additional water. Factors that aid its growth are a well drained soil, high temperatures, and low altitudes. Soil that is too fertile can hinder castor's yield. Cross pollination of castor plants requires that different varieties be grown a minimum of 30 m. apart. Approximately 160 days are required after planting before the first ripe seed is suitable for picking. Because the seeds do not mature at the same time, multiple pickings are necessary for maximum yields. In general, castor is an undemanding and easily grown crop.

Castor oil is unique among all other vegetable oils. It has the highest density and highest viscosity, remains liquid at cold temperatures, is resistant to heat, yet burns well without significant residue.

§ 2 *Modern Production*

Nearly 900,000 m. tons of castor seed are harvested annually throughout the world. Brazil produces about 350,000 m. tons and India 150,000 m. tons;[5] other South American and Asian countries produce most of the rest. Castor grows wild in many countries, including Italy, Egypt, Palestine, and the Mesopotamian countries.[6] Approximately one-half of Brazil's annual yield is harvested from wild plants.

Machines for harvesting castor seed are not often used today because of the varying stages of maturity of the seeds and because of frequent damage to the seeds. Most common is weekly manual

[4] In addition to Weiss, *Castor, Sesame, and Safflower*, pp. 69–78, Eckey, *Fats and Oils*, pp. 745–46, and Vaughan, *Oil Seeds*, p. 201, see E. P. Wright's analysis given in Grenfell, *Revenue Laws*, p. 125: 49–51% oil. Janson, *Castor Oil Production*, p. 2, says that though wild castor seeds may be smaller in size, their oil content is likely to be over 45%.

[5] Weiss, *Castor, Sesame, and Safflower*, pp. 10–44; Janson, *Castor Oil Production*, p. 1; and *FAO 1977 Production Yearbook*, p. 127.

[6] *EB*[11] 5.482; Post, *Flora*, ii p. 509; Thompson, *Assyrian Botany*, p. 132; Dawson, "Studies," p. 69; Lucas, *Egyptian Materials*, p. 332; Schnebel, *Landwirtschaft*, p. 200; Blackman and Fairman, "Myth of Horus," p. 10 n.b.: " . . . castor-oil bushes grow thickly on the banks of the Nile in Lower Nubia at the present day."

harvesting by means of a can with a V-shaped notch cut into its side; the branch of the plant is pulled across the notch, causing the pods to fall into the can. After harvesting, mechanical hullers that can process up to 10 tons of castor per hour separate the seeds from the hulls and shells. Here special care is necessary to prevent breathing the poisonous dust. Seeds then can be stored for long periods of time, since there is no loss in oil quality as long as the skin is not punctured.

Preparation for extraction of oil from the seeds requires separating out any foreign material and drying. Since 5–7% of the seed is moisture, the seeds are dried by means of hot air to keep the moisture out of the oil. The oil is then extracted first by pressing, whether cold, hydraulic, or screw pressing, and then by solvent extraction. At least 12% of the meal left after pressing is oil, which is extracted by adding hexane or ethyl alcohol. The final stage is refining, where the oil is bleached and purified, leaving an almost colorless, slightly yellowish oil. World castor oil production for 1975 was estimated at 340,000 m. tons.[7]

In modern times, castor oil is best known for its medicinal properties as a purgative, but newer medications have largely replaced that use.[8] It is now most important in the cosmetic industry as a softening agent, though it also has applications in hydraulic fluids, in lubricants for heavy machinery,[9] and in the manufacture of resins, textiles, paints, and varnishes. In India, it is still used for illumination.[10] The castor meal is most often used as an organic fertilizer, though it is possible to detoxify the meal and use it in animal feeds. In some areas of Nigeria, castor seeds are eaten by humans after detoxifying them by fermentation.[11] It is reported that in Egypt[12] and Algeria,[13] one castor seed is eaten annually by the women as a birth control measure.

[7] U.S. Dept. of Agriculture, "Statistics," p. 138. For years betwen 1935 and 1958, see Emory/Wolf, "Study of Practices," p. 2.

[8] Martindale, *Pharmacopoeia*, p. 1030.

[9] Swern, *Oil and Fat Products*, pp. 356, 77; Weiss, *Castor, Sesame, and Safflower*, pp. 286–308.

[10] Kirschenbauer, *Fats and Oils*, p. 166.

[11] Vaughan, *Oil Seeds*, p. 75; cf. the numerous medical uses in Boulos, *Medicinal Plants*, p. 86.

[12] W. Blackman and R. R. Marett, *The Fellahin of Upper Egypt* (London: George Harrap, 1927), p. 107.

[13] Hilton-Simpson, *Arab Medicine*, p. 90; cf. Weiss, *Castor, Sesame, and Safflower*, pp. 5–9.

§ 3 *Literary Sources*

References to castor in Greek and Latin literature are usually in connection with Egypt or with medicinal uses of castor. They are in a few cases extensive enough to demonstrate a similar description to modern castor plants and to reveal a familiarity throughout the Mediterranean world with its properties.

Herodotus — Va

2.94 — The Egyptians use an oil from σιλλικύπριον[14] which grows wild in Greece; it is called κίκι. They plant it on the banks of their rivers and lakes where it produces much fruit. The oil is thick[15] and is as suitable as olive oil for lighting though it smells bad.

Theophrastus *HP* — IV–IIIa

1.101 (cf. (3.18.7) — The leaves of the castor plant are curved when new, but angular when older.[16]

Diodorus Siculus — Ia

1.34.11 — Instead of olive oil for lighting, the Egyptians use κίκι.

Strabo — Ia

17.2.5 — Castor seed is pressed to obtain its oil for use in lamps.

Dioscorides — Ip

4.161(164) — Some people call castor wild sesame, others σέσελι κύπριον, and others κροτών because it looks like the insect called κροτών [tick]. The castor plant grows to the size of a small tree and has broad leaves like a plane tree. Its stems are hollow, its fruit spiny and in clusters. Castor is not edible but the oil pressed from it is used in lamps and as a lotion.

1.32(38) — The Egyptians process large quantities of castor for its oil.

[14] This term apparently is a form of σέσελι κύπριον which Dioscorides identifies with castor, see below; LSJ, pp. 1591 col. 2 and 1598 col. 2.

[15] Perhaps quality or quantity and not viscosity is the point (πίων = rich, abundant); but see Pliny *Nat.* 23:41 (83) *pinguitudinem.*

[16] Pliny (*Nat.* 16.25 [85]) makes the same observation.

Pliny *Nat.* Ip

15.7(25) Castor abounds in Egypt; some people call it *croton*, others wild sesame, but our people call it *ricinus* because it resembles the tick. Castor grows rapidly to the height of a small tree both in Egypt and Spain. It is used in lamps but it is not good for food.

23.41(83) Light from castor oil is dim because the oil is too thick.

Though the authors from the second to the seventh centuries also indicate that castor is an oil used in Egypt and is called *kiki*,[17] these later statements may be dependent on earlier sources and do not necessarily reflect the situation contemporary with their writing.[18]

Medical formulas[19] prescribe castor for a variety of purposes: a paste of ground up castor seeds is a cure for warts, irritation of the eyes, unwanted hair, etc.; castor leaves cooked in wine cure inflammations; castor oil purges the stomach, cures muscle stiffness, earaches, and burns, and makes hair grow.

Modern and ancient descriptions of castor are alike in many ways, leaving no doubt of the identification: the height of the plant is like a small tree; the leaves are very large and pointed; the stems are hollow; the seed pods are spiny and are encased in an outer hull and an inner shell; the seeds are not edible; the odor is offensive; the oil is thick, burns well, and is a purgative.

The preceding references to castor in Graeco-Roman literature lead to several conclusions:

1. *Kiki* is the Egyptian term for an oil that the Greeks knew as *kroton* and the Romans as *ricinus*.
2. Castor was cultivated in Egypt, pressed or boiled to obtain its oil, and used in lamps.
3. Castor grew wild in Greece and Spain, but it was generally not cultivated nor was its oil used in lamps.
4. Castor was a common ingredient in medical prescriptions.

[17] Galen, Oribasius, Hesychius, Georgius Choeroboscus, and Paulus Aegineta; citations and quotations are in *TGL* V pp. 1550 col. 2, 1551 col. 1, 2004 col. 2, 2005 col. 1; see also LSJ.

[18] Cf. below n. 24 regarding *kiki* in Coptic.

[19] In particular: Dioscorides 1.32(38); 4.161 (164); Pliny *Nat.* 23.41 (83); Hippocrates 7.358; 8.386; Galen 12.26; Paulus Aegineta 7.3; Celsus 5.19.26.

§ 4 *Ancient Civilizations*

In addition to the notices of castor in Classical literature, Assyrian texts reveal that castor was commonly used as a drug, since it appears in numerous medical prescriptions.[20] There is also an apparent reference to castor in the Hebrew scriptures: the plant that provided shade for Jonah (4.6–10) is קיקיון. That has been linked to the Assyrian *ḳûḳânîtum*[21] and to the Egyptian *kȝkȝ*[22] (the Septuagint, however, reads κολοκύνθη).[23] The Jewish military colony at Elephantine used castor extensively as their Aramaic texts reveal. Here they borrow the Egyptian term *tgm* and transcribe it תקם.[24] As is true of the Demotic texts, תקם occurs frequently in marriage contracts from Elephantine.[25] However, in the Talmud (*Shabbath* 21a), there is uncertainty about the identification of *ḳiḳ*.

From Egypt there is much evidence of the use of castor. This was demonstrated above (ch. 1 § 7, pp. 32–34) from Demotic texts, but castor in Egypt significantly antedates the beginning of the Demotic script.[26]

Castor seeds from an archaeological excavation of a grave have been dated to the Badarian period (Predynastic, c. 5000 B.C.).[27] Numerous finds of seeds from Pharaonic Egypt that are now in the museums of Berlin, Vienna, and Paris provide additional examples of castor seed.[28] Seeds found in a tomb at Thebes, though once thought to be ancient castor, were probably dropped by modern Arabs.[29]

[20] R. C. Thompson, *The Assyrian Herbal* (London: Luzac, 1924), pp. 190–191; Thompson, *Assyrian Botany,* p. 132 *et passim.*

[21] F. Delitzsch, *The Hebrew Language Viewed in the Light of Assyrian Research* (London: Williams and Norgate, 1883), p. 24; BDB, p. 884.

[22] H. Farzat, "Encore sur le mot *TQM* dans les documents Araméens d'Éléphantine," *Semitica* 17 (1967):79; Porten, *Archives,* p. 92.

[23] Dawson therefore says it is *Cucurbita pepo,* climbing gourd; "Studies," p. 70.

[24] Keimer, "Naturgeschichte," 100–104; P. Grelot, "L'huile de ricin à Éléphantine," *Semitica* 14 (1964):64–70; A. Dupont-Sommer, "Note sur le mot *TQM* dans les ostraca araméens d'Éléphantine," *Semitica* 14 (1964):71–72; see above n. 22. For other Semitic terms for castor, see Löw, *Pflanzennamen,* pp. 352–54. For the Coptic terms, see W. E. Crum, *A Coptic Dictionary* (Oxford: Clarendon, 1939), p. 466, and Till, *Arzneikunde,* p. 89.

[25] Porten, *Archives,* pp. 92–93.

[26] The best surveys of castor in antiquity are: Woenig, *Pflanzen,* pp. 337–40, Keimer, *Gartenpflanzen,* pp. 70–73, and H. Stadler, "Ricinus," *RE* I A (1914), cols. 800–802.

[27] G. Brunton and G. Caton-Thompson, *The Badarian Civilization* (London: British School of Archaeology in Egypt, 1928), p. 38.

[28] Keimer, *Gartenpflanzen,* p. 71.

[29] So Schweinfurth personally reported to Keimer; *ibid.* p. 71, 119 n. 17.

Castor seeds have also been found from the Roman period (second or third century) in a cemetery in Hawara in the Fayum.[30] From Pharaonic times, containers made from the castor plant are extant,[31] as well as tomb paintings picturing plants in stylized form. The latter, however, are insecurely identified with castor.[32] Oil stains on mats and linen found in tombs were first chemically analyzed to be castor oil, but more recent studies have shown them to be from animal fat.[33]

Castor is likewise mentioned in the written documents of ancient Egypt. The Demotic term for castor, *tgm* (above, ch. 1 § 7, pp. 32–33), has an equivalent that appears in much earlier Hieroglyphic and Hieratic texts. Most common in the medical papyri, castor in the form of its seeds, roots, and oil is included in incantations, spells, and prescriptions:[34] castor in combination with other seeds, when burned, will drive away evil spirits; castor, crushed and mixed with honey, relieves pain when applied as a bandage; castor seed mixed with grease makes the hair of women grow.[35]

In addition to the medical papyri, an inscription of the sixth century B.C. records the dedication to a temple of castor oil for illumination.[36] The term *k3k3* also means castor and, though occurring less frequently, appears in medical formulas.

The use of products of the castor plant was therefore very common in pre-Ptolemaic Egypt as the material and written remains show.

§ 5 *Occurrences in Papyri*

The references to castor are listed in appendix D. The list of documents includes numerous accounts, receipts, and letters; several memoranda and petitions; and an occasional will, land survey, complaint, contract, and list. Two archives figure prominently: the Zenon collection mentions castor frequently, along with many other agricultural products; and the texts from the Serapeum at Memphis are particularly concerned with castor oil.

[30] W. M. F. Petrie, *Hawara, Biahmu, and Arsinoë* (London: Field and Tuer, 1889), p. 51.

[31] W. C. Hayes, *The Scepter of Egypt,* Part II: *The Hyksos Period and the New Kingdom* (Cambridge: Harvard University, 1959), p. 191.

[32] Keimer, *Gartenpflanzen,* p. 71; Dawson, "Studies," p. 63.

[33] Lucas, *Egyptian Materials,* p. 328.

[34] Dawson, "Studies," pp. 52–57, 62; Lucas, *Egyptian Materials,* p. 332.

[35] Examples from the Ebers papyrus as reported in Dawson, "Studies," pp. 53–54. See Grapow/Deines, *Wörterbuch,* p. 583, and Kamal, *Pharaonic Medicine,* pp. 89–90.

[36] Dawson, "Studies," p. 62.

§ 6 *Terminology*

Throughout ***P.Rev.*** κροτών and κίκι are consistently differentiated, the first referring to the seed or plant and the second to the oil. The same distinction generally is made in the other papyri, but there are several exceptions. The term κροτών is never used for oil, but κίκι refers to seed or plant in several papyri: ***P.Cair.Zen.*** II 59176.34; 59243.14; IV 59552.3; 59629.2–3, 5; ***P.Lond.*** VII 1959.2, 6; 2061.7; 2164.6; ***PSI*** V 500.5; and ***SB*** VIII 9667.3. Similar freedom in the use of κίκι is attested outside the papyri, as in Strabo 17.2.5 where it is used for seed, and Galen 19.743 where it is used for the plant: κροτῶνες κίκεως.

§ 7 *Plant Description*

One text indicates something of the nature of the castor plants. Unfortunately the fragmentary nature of ***P.Cair.Zen.*** II 59176 hides the immediate context of ξύλα κίκιος, although it is certain that the document is an account of daily expenses. The papyrus, however, does use the term ξύλον for a castor plant. As noted above, castor plants are known today to reach the size of a small tree, and Dioscorides and Pliny noted the fact in antiquity as well. The same is true for Ptolemaic Egypt as this papyrus suggests. Castor, therefore, was apparently a perennial. Modern varieties are perennial, and the evidence of castor plants as large as trees indicates that castor was perennial in antiquity also.

§ 8 *Agricultural Details*

Several papyri speak of planting castor and one suggests that castor was planted by setting out small plants or shoots: ***PSI*** V 499: τὰ φυτά have been sent εἰς [τ]ὴν φυτε[ί]αν τοῦ κροτῶνος. Four other papyri verify this form of planting castor:

PSI	V	500	κίκιος φυτεία
P.Cair.Zen.	II	59243	κίκι φυτεῦσαι
	IV	59552	κίκιος φυτείαν
P.Lond.	VII	1959	φυṭẹί̣α̣ṿ κίκιος

On the other hand, ***P.Cair.Zen.*** II 59292 and ***P.Lond.*** VII 1994 and 1995 specify some art. of castor for seeds (εἰς σπέρμα). Also, ***P.Lond.*** VII 2061 is a letter that refers to the addressee sowing castor: κίκι ἔσπιρας. Likewise, Herodotus 2.94 and Strabo 17.2.5 speak of sowing castor. Therefore, castor was started both from seed and from shoots. In modern practice, castor seed is only sown, either by hand or mechanically.[37]

It is probable that by Ptolemaic times castor would have grown wild in Egypt, since that is true today and since castor had been growing in Egypt before the Ptolemaic period. No reference to picking wild castor, however, has been found and the absence in ***P.Rev.*** of any government restriction may argue against it. Only one text hints at wild castor. ***P.Cair.Zen.*** 59635 is a letter reporting that castor grows in salty soil (the statement that the salt was three fingers thick was crossed out): τὸ δὲ λοιπὸν ἁλμυρίδα [[ἐπὶ δακτύλους τρεῖς τὸ πάχος]], φάσιν δέ ἀναβήσεσθαι κροτῶνα. As indicated above, castor grows well in less fertile places.

The season for planting castor is not clearly revealed in the papyri, though it would be expected to conform to the flooding of the Nile. Four papyri that mention planting castor give specific dates which suggest a spring rather than a fall planting:

P.Cair.Zen.	II	59243	(3 February)
	IV	59552	(14 March)[38]
P.Lond.	VII	1959	(14 March)
PSI	V	499	(22 February)
		500	(7 July)

Schnebel suggests that there was not a fixed time for planting.[39]

The cost of having a field planted in castor is given in two papyri of 256 B.C., though the quantity planted is not stated: ***P.Cair.Zen.*** IV 59552, 5 dr. 2 triobols; and ***P.Lond.*** VII 1959, 5 dr. The only mention of the care of castor plants after they are planted is ***P.Lond.*** VII 2164 where some young workers are weeding the castor: εἰς παιδάρια τὰ τὸ κίκι βοτανίζοντα. No evidence has been found for fields of castor that could be used repeatedly (since castor is a perennial), nor has evidence been found for how castor was harvested.

§ 9 *Amounts Cultivated*

The nomes assigned to grow castor as given in ***P.Rev.*** 60–72[40] are only eight (in contrast to sesame, which is assigned to all twenty-

[37] Janson, *Castor Oil Production*, p. 2.

[38] ***SB*** VIII 9667 is the other half of this document, recording the same information; ***P.Lond.*** VII 1959 concerns the same transaction. ***P.Lond.*** VII 2061 may reflect a spring planting of castor.

[39] Schnebel, *Landwirtschaft*, p. 201.

[40] See appendix A.

TABLE 2 *Castor Assignments in **P.Rev.** 60–72*

	For Use Within the Nome	For Other Nomes	For Alexandria
Saite with Naucratis	11,433 ⅔ arou.		10,666 ⅓ arou.
Prosopite	2,000 arou.		13,600 art.
Busirite	[. . .]	[. . .]	
Athribite	3,760 arou.	3,000 art.	
Letopolite	550	1,250 arou.	
Arsinoite	[. . .]	[. . .]	
Cynopolite	[. . .]		
Thebaid	11,820 arou.		9,067 arou.

four,[41] except the Memphite). Those eight nomes would be expected to produce large amounts of castor in order to supply themselves plus the sixteen nomes producing none; especially so when the number of art. to be supplied to twelve of those sixteen nomes not producing any castor is 72,220 (the figures for the other four are lost), and when the eight nomes are to supply Alexandria in addition to this. Only four nomes, however, are to provide castor for the sixteen not growing any. The eight nomes and their obligations are listed in table 2 (for the 72,220 art. that the other nomes are to receive, see appendix A).

Only three of the eight nomes to produce castor—the Arsinoite, Thebaid, and Cynopolite—are outside the Delta; and though castor was to be provided for all the nomes plus Alexandria, only Saite with Naucratis and Thebaid are designated to plant large amounts of land in castor. Perhaps the limitation of castor production to a few nomes is not surprising given the evidence that castor was widely used in Egypt for centuries before this, given the statement of Herodotus (2.94) that castor was generally planted on the banks of rivers and lakes, and given that castor was a perennial. A few nomes, therefore, that had an established castor crop could produce large amounts of castor without new plantings.

In addition to ***P.Rev.***, a number of other third century papyri report various amounts of castor. The references to arou. planted are all from the Arsinoite nome:

[41] Memphis and the Memphite nome are listed separately; I count them as one. I also count the Thebaid as a nome for a total of twenty-four nomes; see appendix A.

5 arou.[42]	***P.Cair.Zen.*** IV 59552	256a
5 arou.	***SB*** VIII 9667	256a
20 arou.	***P.Cair.Zen.*** II 59243	252a
130 arou.	***P.Petr.*** II 39a	IIIa
55 arou.	III 75	234–33a
120 arou.	III 88	IIIa

P.Lond. VII 2061 may belong in this list. It is a letter to Zenon that first mentions the irrigation of 500 arou. Later in the letter it refers to the land planted with castor. Whether the land planted in castor is the 500 arou. is not clear, but the amount is large compared to the figures in the list above. ***P.Petr.*** III 75, a report for the Arsinoite nome twenty-five years after ***P.Rev.***, records 55 arou. in castor out of a total of 180,014 $^{29}/_{32}$ arou. in twelve different crops (the papyrus lists thirteen crops, but the figure for κύαμος is lost). By comparison, 134,315½ arou. were sown in πυρός. This report, however, is of land planted by 18 January in fall/winter crops. This small amount of land planted with castor does not then reflect what would have been planted in the spring.

P.Cair.Zen. II 59292, ***P.Lond.*** VII 1994 and 1995 record deliveries of castor seed to Heracleides. The total receipts for three years were only 21 art. of castor seed. ***P.Petr.*** II 39a.2, 5, 9 and III 88.2, 5–6, in addition to listing arou. planted, also give amounts of castor for those arou.

for 40 arou.,		3 (measures) of castor				
" 20	"	1½	"	"	"
" 50	"	3½	"	"	"
" 20	"	1½	"	"	"
" 20	"	1½	"	"	"
" 100	"	7½	"	"	"

It has been assumed[43] that this is an account of the number of art. of seed given out for sowing a particular number of arou.; the ratio is ¾ art. for 10 arou. If castor was being planted from shoots instead, these figures would refer to quantities of shoots.

Only two records of harvested castor are known: from Phthemphouth, ***P.Tebt.*** III pt. II 845 (the highest figures in this list of various amounts are 295¼ and 391 art., with a recorded total harvest of less than 1,000 art.) and from the Fayum, ***Chrest.Wilck.*** 304 (a total of 1,667 art.).

[42] ***P.Cair.Zen.*** IV 59552 and ***SB*** VIII 9667 refer to the same 5 arou.

[43] By the editors of ***P.Petr.*** (III p. 224) and Jean Bingen, "Papyrologica: ***P.Oxy.*** 2191, ***P.Heid.*** 244, ***P.Petrie*** II 39 (a)," *Chronique d'Égypte*, XXXIX (1964), p. 172.

P.Cair.Zen. III 59326.3 and IV 59787.8–9, 43, 45, 52 are accounts of money spent for various items in which castor is included. Expenditures of 100 dr. for castor are mentioned four times in the two documents. In addition, a marginal note at line 52 of 59787 gives a quantity of 100 art. purchased for 100 dr.

P.Lond. VII 1991.3, 93 records the amounts of castor stored in the granaries on Apollonius's estate in 251 B.C. The extant figures are: $61\frac{1}{2}$ and $84\frac{1}{2}$ art. ***P.Cair.Zen.*** IV 59717.7 mentions $179\frac{1}{3}$ art. of castor on hand.

P.Col. III 40.5 is a receipt for $100\frac{2}{3}$ art. of castor turned over by an agent of Zenon. ***PSI*** IV 358.5, 18, 31 is also a receipt for a similar amount of castor, 100 art.

From the second and first centuries are references to castor in various amounts: ***P.Tebt.*** III pt. II 829 mentions 20 arou. of castor planted and a rent of $4\frac{45}{48}$ art. per arou.; ***BGU*** VI 1217 lists a rent payment of $1800\frac{1}{2}$ art. of castor; ***O.Bodl.*** I 171, ***O.Wilck.*** II 727, 729, 737, 741, and 743 are a group of ostraca that record amounts of castor for Thebes, from 2 to 23 art.

§ 10 *Oil Production*

Though ***P.Rev.*** regulates factories, its only information regarding amounts processed is inferred in columns 60–72. As indicated above, eight nomes produced amounts of castor for their own use, for Alexandria, and 72,220 art. for twelve nomes producing none. With each of the quantities for these twelve nomes is the phrase ὃν δεῖ κατεργασθῆναι ἐν τῆι ὠνῆι, indicating that the castor must be processed into oil under the usual auspices of the tax contract. It is apparent that the seed was to be processed in all the nomes, for seed and not oil was to be supplied, and a factory is specifically mentioned at Alexandria (***P.Rev.*** 58.6 = 60.12).

The specific references to factories in other papyri represent the Arsinoite, Aphroditopolite, and Thebaid: ***P.Cair.Zen.*** III 59412.6; ***P.Tebt.*** III pt. II 844.10; ***O.Wilck.*** II 737.2; 741.2; 743.2. In the first two of these papyri, sesame and castor are reported at the same factory. Presumably, separate presses were used for the two products due to the toxic nature of castor. The second papyrus states that $22\frac{1}{6}$ art. were on hand but were not processed.

§ 11 *Yield Possibilities*

The scattered records of amounts of castor planted and harvested unfortunately leave out any data on the yield of castor, and no evidence from antiquity resolves the question. Modern production levels are the only source of information,[44] but they must be viewed with considerable uncertainty: although castor is said to grow wild in Egypt,[45] no landholder or agricultural worker reported castor as a crop in the modern agricultural censuses of Egypt.[46] Thus, alternative countries must be selected to provide a standard. In addition, differences in cultivation, variety of plants, and cultivation may affect the comparison of ancient and modern yields.

The following countries were selected as representative of yield possibilities and as having some affinities with Egypt (quantities listed are kg. per ha.):[47]

	1948–52	1952–56	1961	1962	1969–71	1976	1977
India	200	200	230	220	304	381	353
Iran	570	590	610	590	529	1000	1000
Italy	1280	1180	1200	1670			
Libya	1580	1780	2000	2190	1060	1833	1833
Mexico	350	380	680	680	500	578	556
Sudan	700	700	750	790	1020	833	927

It should be noted that since 1960, yields have increased for several countries because of improvements in cultivation techniques and in varieties of castor; however, during the previous fifty years,[48] the levels remained constant. Therefore, for the purpose of this comparison earlier figures will be used rather than the more recent.[49] India, though with a low average yield, demonstrates wide diversity within the country for yield statistics: eleven states report castor production for the years 1956–60 and their yields vary from 122 to 837 kg. per ha.[50] Unlike Egypt, India has varied elevations, tem-

[44] Another attempt to determine ancient yields of castor using modern production levels was made by John Wickersham, though with somewhat different figures: J. Wickersham, "The Financial Prospects of Ptolemaic Oilmen," *BASP* VII (1970):45–51.

[45] Above, § 1, p. 35.

[46] E.g., *Report on the 1960 World Census of Agriculture: Census Results by Countries*, Food and Agriculture Organization of the United Nations (Rome, 1966), vol. I. pt. A, pp. 206–23.

[47] From vols. 20 (1966) and 31 (1977) of *Production Yearbook*, Food and Agriculture Organization of the United Nations (Rome, 1967 and 78).

[48] Most clearly seen for India; E. L. Burtis, "The Fats and Oils Economy of India," Foreign Agricultural Service, U.S. Department of Agriculture (July, 1960), p. 3.

[49] The earliest figures available for castor in a worldwide report are from the 1966 *Yearbook* (see n. 47 above) which gives figures back to 1948; figures for India are available back to 1910; see n. 48 above.

[50] *Oil Seeds in India: 1956–57 to 1960–61* (India: Ministry of Food and Agriculture, 1962), p. 31.

peratures, rainfall, and soil, which may account for this disparity in yields.

On account of the wide variance in modern yield results and the uncertainty of comparisons with ancient Egypt, three calculations will be made, one high, one medium, and one low.

Selecting the two countries closest to Egypt, Libya and Sudan, and taking the average of their yields for the years 1948–56, an approximation can be calculated from modern times for yields expected in Ptolemaic Egypt:[51]

Average yield (Libya and Sudan) = 1190 kg. per ha.

0.23 ha. = 1 arou.[52]
1 arou. = 273.7 kg.

Converting a yield of 273.7 kg. per arou. to volume results in:

1 bu. of castor = 18.6 kg.[53]
1 art. = 1.22 bu.[54]
1 art. = 22.7 kg.

Therefore, 273.7 kg. per arou. is equivalent to 12 art. per arou.

On the other hand, using the average of Iran and Sudan[55] for 1948–56, 640 kg. per ha., and performing the same calculations as above, the result is a a yield of 6.5 art. per arou. Using the average of India and Mexico,[56] 282.5, results in a yield of only 2.9 art. per arou.

What is available in the papyri to guide the choice from the above three possibilities is unfortunately limited. ***P.Rev.*** 60–72 may be of some help:[57] the number of art. of castor to be received by the

[51] Note that the average of Libya and Sudan is very close to the yield reported by Italy.

[52] Above, ch. 1 § 3, pp. 9–10.

[53] U.S. Department of Agriculture, "Statistics," pp. v–vi, give a bu. of castor at 41 U.S. lbs. or 18.6 kg. Jamieson, *Fats and Oils,* p. 47 gives 46 lbs. per bu.

[54] Above, ch. 1 § 3, pp. 9–10.

[55] I use Sudan twice, first averaged with Libya, and now with Iran, because it has many likenesses to Egypt, especially the Nile.

[56] I include Mexico on the example of Wickersham, "Ptolemaic Oilmen," pp. 49–50; he selected Mexico alone as a guide to castor yield on the criterion that Egypt and Mexico currently have similar sesame yields (the assumption that two countries that have similar yields in one agricultural product will have similar yields in all others is invalid). Wickersham furthermore used Mexico's yield for 1961 and 1962 (680 kg. per ha.), ignoring that the yield for 1948–56 was signficantly lower (365 kg. per ha.).

[57] See appendix A.

non-castor-producing nomes ranged from 2,120 for the Memphite to 12,900 for the Hermopolite. Taking the five figures extant of the number of arou. that nomes were to plant for themselves, and multiplying them by the proposed yields, produces some striking results:

	arou.	× 12	× 6.5	× 2.9
Saite with Naucratis	11,433⅔	137,196	74,319	33,158
Prosopite	2,000	24,000	13,000	5,800
Athribite	3,760	45,120	24,440	10,904
Letopolite	550	6,600	3,575	1,595
Thebaid	11,820	141,840	76,830	34,278

Add to this the 20,983⅓ arou. (in addition to the 16,600 art.) for Alexandria:

arou.	× 12	× 6.5	× 2.9
20,983⅓	251,796	136,392	60,852

Certainly the higher yields result in quantities of castor for the nomes producing them that are disproportionate to those amounts for non-producing nomes (12,900 art. is the highest). It could perhaps be argued that the nomes producing castor were in fact growing it because they had very large needs for it, while those not assigned castor production would only receive small quantities. That seems unlikely in light of the rest of the oil regulations of ***P.Rev.***

The only other control for choosing an approximate castor yield is ***P.Tebt.*** III pt. II 829, a land survey of the second century B.C. that reports rents per arou. for various crops. The rent was the common 4 45/48 art.[58] per arou. for most of the crops—castor, barley, etc. (the papyrus clearly says that the rent was set κατὰ τὴν ἀρετήν [line 24], which must mean that the yield was sufficiently plentiful to allow a profit while still paying in kind 4 45/48 art. per arou.). Yield/rent ratios are not well known for Ptolemaic Egypt; for wheat the evidence suggests a yield of 10 art. and a rent of 4 art. per arou.[59] Thus, for castor, a yield of 2.9 is too low to pay a rent and 12 exorbitant for the

[58] For other examples of the same rent, see ***P.Tebt.*** I 61–66, also IIa.

[59] Crawford, *Kerkeosiris*, pp. 125–27. A. Segré, "Note sull'economia dell'Egitto ellenistico nell età tolemaica," *Bulletin de la Société d'Archéologie d'Alexandrie* XXIX (1934), pp. 257–305, uses an average of a twelve-fold return for wheat; Whitney

P.Rev. tables. The best approximation, then, of castor yield for Ptolemaic Egypt is 6.5 art. per arou. (the figures for Libya and Sudan for 1948–56).

Evidence on the rate of seed sown per arou. comes from ***P.Rev.*** 41.14–18 and two other papyri: 2 dr. are allotted per arou. for seed. Since column 39 prescribes a value of 4 dr. per art. for castor, the intended planting ratio was ½ art. per arou. ***P.Petr.*** II 39a.2, 5, 9 and III 88.2, 5–6 may be allotments of seed for sowing. The ratio is ¾ art. for 10 arou. or 3 choi. per arou. With the uncertainty of interpretation of these last two documents, the planting ratio suggested in ***P.Rev.*** is probably more accurate.

For the ratio of oil from seed, again no information exists in the papyri. Modern figures are: 100 lbs. of castor seed yield 5 gal. of oil by screw press.[60]

100 lbs. castor seed = 45.5 kg.	5 gal. oil = 18.9 l.
22.7 kg. = 1 art.	39.39 l. = 1 metr.
100 lbs. castor seed = 2 art.	5 gal. oil = 0.49 metr.

Thus, from 2 art. of castor, 0.49 metr. of oil could be pressed. It would then require 4 art. of castor to produce 1 metr. If the yield of 6.5 art. per arou. is correct, then from the seed of one arou., 1.6 metr. of oil could be expected (disallowing taxes and seed for new crops).

§ 12 *Amounts of Oil*

Several papyri provide interesting data about how much oil was used on a daily basis as well as how much was spent to purchase oil. ***P.Corn.*** 1,[61] especially valuable here, is an account, for a period of fifty-nine days, of assignments of castor oil to various departments of the retinue of Apollonius on a tour of Egypt. The total amount allotted for this period was 336⅛ kotylai,[62] varying from 5 to 11¼ kotylai per day. Examples of the departments and of the castor oil allotted per day are: for an office of accounting, one kotyle; for a bakery, ½ kotyle; for a storeroom, ¼ kotyle; for a horse stable, ¼

Scofield Bagnall, "The Archive of Laches: Prosperous Farmers of the Fayum in the Second Century" (Ph.D. diss., Duke University, 1974), p. 113, n. 12, pp. 119–20, found rents paid of 10–14 art. per arou., which must be lower than what the yield per arou. would be.

60 *EB*[11] 5.482.

61 The original edition of this papyrus: W. L. Westermann, "Account of Lamp Oil from the Estate of Apollonius," *Classical Philology,* XIX (1924), pp. 229–60.

62 Above, ch. 1 § 3, pp. 9–10 on the kotyle.

kotyle. These figures varied according to anticipated need. For example, when the bakery was preparing for a festival, the allotment was increased from the normal ½ kotyle to 1½ kotylai and then to 2½ kotylai per day. Samuel,[63] attempting to find the burning life of a kotyle of castor oil, burned one-half pint of castor oil in each of two ancient lamps using wicks of braided linen and hemp. He found that a kotyle burned in one lamp 14½ hours and in another 18½ hours. Assuming that the oil was not needed for work in daylight hours, this means that considerable work was done after sunset and before sunrise; for example, if two people normally worked in the bakery, using one lamp each and receiving ½ kotyle of oil per day, each worked during 3½–4½ hours of darkness; but in preparation for a festival, when the bakery received 2½ kotylai per day, ten people could have worked 3½–4½ hours at night.

Daily allotments of castor oil similar to ***P.Corn.*** 1 are common in other papyri: ***P.Cair.Zen.*** IV 59689.8; V 59809.5–6, 8; ***P.Col.*** III 37.21; ***P.Petr.*** II 25a.14; b.16; c.1; ***UPZ*** II 204.3–4.

Other papyri that indicate amounts of castor oil are: ***P.Mich.*** I 72.5, a request for three ch.; ***PSI*** IV 349.2, a request for 60 metr.; and ***UPZ*** 17–120, a request for 1 metr. per year.

§ 13 *Seed and Oil Prices*

P.Rev. 39 gives a price for castor seed at 4 dr. per 30 choi. art. The only other indication of castor seed price is the marginal note in ***P.Cair.Zen.*** IV 59787 mentioning an expenditure of 100 dr. for 100 art.

Expenditures of ¼–4 ob. per day for castor oil are recorded in several third century papyri: ***P.Cair.Zen.*** IV 59704.5, 14, 17, 30; 59705.17, 29, 38; ***P.Col.*** III 37.21; IV 108.3; ***P.Hib.*** I 121.17, 21, 29, 36, 44; and ***UPZ*** II 158a.36, 62, 64, 78, 102, 124, 142. Other expenditures are: ***BGU*** VII 1519.13(III), 100 dr.; ***P.Mil.*** II 27ii.45(158), 1,380 dr.; ***P.Sorb.*** 16.12(257), 13 dr.; ***P.Tebt.*** II 570(I), 50 and 80 dr.; and III pt. II 865.2, 23, 27(III), 334 dr. in the course of four months.

The price for castor oil specified in ***P.Rev.*** was 48 dr. per metr. (= 4 dr. per ch. or 2 ob. per kotyle). Two papyri contemporary with ***P.Rev.*** use the same pricing: ***P.Col.*** III 21.4, 1 ch. lost, worth 4 dr.; and 37.21, 2 kotylai purchased for 4 ob. Two second century papyri, however, report several purchases of 1 kotyle for 28, 50, and 60 dr. per kotyle, indicative of a severe inflation:[64] ***P.Tebt.*** III 885; and 891.

[63] Alan E. Samuel, "Illumination by Castor Oil—P.Cornell 1," *Bulletin of the American Society of Papyrologists* I (1963), pp. 32–38.

[64] See T. Reekmans, "The Ptolemaic Copper Inflation," *Studia Hellenistica* 7: *Ptolemaica* (1951) 61–119.

A comparison of daily expenditures for castor oil and other necessities is possible in four papyri. In ***P.Cair.Zen.*** IV 59704; 59705; ***P.Hib.*** I 121, and ***UPZ*** II 158a, 1/4–1/2 ob. per day is common for wood, radishes, cabbage, salt, hyssop, hot water, and castor oil. Thus, castor oil was comparable in price to other daily necessities. Expenditures for bread, wine, and fish are generally higher: 1–3 ob.

§ 14 *Uses*

The use of castor oil for lighting in Ptolemaic Egypt is evident from several papyri. ***P.Cair.Zen.*** IV 59689.8 is an account of castor oil to be supplied to a stable, ἐπὶ λύχνον; ***P.Col.*** III 37.21 records an expenditure for castor oil for the preparation of copper pots during the night (presumably having nothing to do with the pots, but providing light for night-time work); ***P.Corn.*** 1 is a daily record of disbursements of castor oil, with the phrases ἐπὶ λύχνον and ἐπὶ λαμπτῆρα occurring frequently; ***P.Petr.*** II 25a.14; b.16; c.1; e.7 is an account of castor oil received εἰς λύχνους or εἰς παράκαυσιν. ***UPZ*** 204.3–4 mentions castor oil for use by troops εἰς τὰς [ν]ῦκτας, and ***P.Cair.Zen.*** IV 59706 concerns castor oil for a journey, perhaps at night.

Castor oil is used with natron as a soap in ***P.Cair.Zen.*** 59304 and ***P.Tebt.*** III pt. I 703.102. It is also furnished to λινεψοί in ***PSI*** IV 349. ***P.Lille*** 58iii.18 (βαλα[νεῖ] κίκι) may also suggest that castor oil is connected with bathing.

§ 15 *Comparisons and Conclusions*

In contrast to sesame oil and olive oil, castor oil was in common use in Egypt before the Ptolemaic period. When the Greek settlers arrived, they found castor oil a poor substitute for olive oil; nevertheless, the Greek and Demotic papyri demonstrate that castor oil continued to be widely used. By Roman times, however, things had changed. Castor is rarely mentioned in Roman papyri as a crop or source of oil. Dioscorides and Pliny continue the common literary references[65] to castor in Egypt, but those reports may be based on the castor plants growing wild. Radish seed and olives became the most common oil sources of Roman Egypt.

Since castor did not need to be planted each year, unlike sesame, an ample supply of castor seed was possible without annual plantings. This may explain why the references to castor in the papyri are most

[65] Below, ch. 3 n. 31, p. 71.

often to oil and less frequently to planting and harvesting seed. Further, according to ***P.Rev.*** 60–72, the nomes in the Delta were assigned the majority of oil seed cultivation. Unfortunately, with little papyrological evidence for the Delta, the castor production there is not well documented.

III

Sesame
(σήσαμον)

§ 1 *Physical Description*[1]

Sesamum indicum L. (Fam. Pedaliaceae) is an erect plant that flourishes in tropical and subtropical regions. Its leaves vary from broad, lobed leaves at the base of the plant to narrow, pointed leaves at the top. The leaves are green but often have reddish-colored veins. It is an annual plant that has the shortest growing season of any oil seed crop; within 85–130 days mature seed is produced. The height of the sesame plant is normally 100–120 cm., though some conditions have produced plants up to 180 cm. After about four weeks of growth, when the plant is about 30 cm. in height, flowers appear in the axil of each leaf, normally one per leaf. As the stem grows higher, more and more flowers appear. In those same locations the fruit is produced which is a narrow capsule 2–3 cm. in length with many seeds in four or eight rows. Numerous varieties of sesame are grown today (30 or more are reported in India). These varieties differ primarily in height, in manner of branching, and in the shape of the capsules. In general, the sesame plant branches out near its base into several long stems, each producing many seed pods.

[1] The most thorough discussion of sesame is Weiss, *Castor, Sesame, and Safflower,* pp. 356–91; see also A. B. Joshi, *Sesamum* (Indian Central Oilseeds Committee, 1961); Eckey, *Fats and Oils,* pp. 741-48; Vaughan, *Oil Seeds,* pp. 201-03; Jamieson, *Fats and Oils,* pp. 236–39; P. Alpin, *Plantes d'Egypte* (Institute français d'archéologie orientale, 1980 reprint of 1640 ed.), pp. 138–41; J. C. Bahl, *The Oilseed Trade of India,* (4th ed.; Bombay: New Book, 1938), pp. 25–26; P. S. George, U. K. Srivastava, B. M. Desai, *The Oilseeds Economy of India: An Analysis of Past Supply and Projections for 1985* (Delhi: MacMillan of India, 1978), pp. 72–94; *Grain and Oilseeds,* part IV, pp. 24-26; Usher, *Dictionary of Plants,* p. 536; Trease and Evans, *Pharmacognosy,* p. 329; Martindale, *Pharmacopoeia,* p. 1034; Bedevian, *Plant Names,* p. 544.

SESAME

Sesame seeds are pear-shaped, flattened on two sides and pointed at one end. They are 2.5–3.0 mm. long and 1.5 mm. wide, weighing between 2.0 and 3.5 gr. per 1,000 seeds. A thin shell surrounding the seed may be removed by soaking and rubbing, but that is unnecessary prior to oil extraction. In color the seeds vary from white to yellow to brown to black. About 50% of sesame seed is oil and about 25% is protein, making it an attractive oil source. Sesame is also desirable because its oil is of good quality.

Sesame grows well on light, sandy soils that would be judged inadequate for many crops. It requires, however, several waterings from rain or irrigation and can only tolerate short periods of drought. It can be grown during any season of the year given a warm climate. Sesame also grows well in dense plantings. Whether sown in carefully spaced rows or hills, or broadcast, the yield is nearly the same (if sesame plants have ample room, they grow numerous stems; if little room, they grow only one). Because of the small size of the seeds, sesame is often easier to sow generously and then thin out the plants after germination.

The uneven ripening of the fruit of the sesame plants and the bursting of the seed capsules as soon as mature creates some special problems for the harvest of sesame. Generally, the entire plant is taken from the fields as soon as the leaves of the plant begin to turn yellow, but while the seed pods are still green. After about one week the leaves fall off the bundled plants, and after about two more weeks all the pods have opened. Then the plants are shaken or beaten lightly on a floor or over a cloth to release the seeds.

§ 2 *Modern Production*

Nearly two million m. tons of sesame seed are harvested annually throughout the world, primarily in African and Asian countries. Approximately 1/5 of that total is grown in China and approximately 1/4 in India. In recent years Egypt has produced about 15,000 m. tons.[2] In 1920 Egypt reported 4,800 ha. planted in sesame from which 3,940 m. tons of sesame seed were crushed to produce 1,790 m. tons of oil.[3] Were it not for the problems of harvesting sesame, making

[2] *FAO 1977 Production Yearbook*, p. 130; K. F. Mattil, "Review and Comparative Analysis of Oilseed Raw Materials and Processes Suitable for the Production of Protein Products for Human Consumption" (New York: United Nations, 1974), p. 3; cf. Emory/Wolf, "Study of Practices," p. 2, and Weiss, *Castor, Sesame, and Safflower*, pp. 319–55.

[3] *Oleaginous Products and Vegetable Oils*, pp. 306–7; cf. E. L. Burtis, "The Fats and Oils Economy of India" (U.S. Dept. of Agriculture, 1960), p. 3.

mechanical harvest difficult, sesame would be much more widely grown today as a commercial crop.[4]

Sesame oil is typically expressed with hydraulic presses in three stages:[5] the first is a cold pressing, the second and third apply increasing amounts of heat and pressure. Normally the oil from the first pressing is ready for use, while that from the second and third pressings must be refined. In some cases, the remaining press cake is subjected to solvent extraction to remove even more oil before it is used in cattle feed or as a fertilizer. Primitive methods of extracting sesame oil are still in use in some parts of Africa and India[6] where either the seed is pounded in a wooden mortar, boiled in water, and the oil skimmed off, or an animal-powered mill presses the seed to extract the oil.

Most of the sesame produced is used in foods: the oil as a component of vegetable shortenings and as a cooking oil, and the seed in bakery and confectionary products. Sesame oil is also used in soaps, in paints, in medicines, and occasionally for lighting.[7] In Algeria, healing practitioners prescribe fumigation in the smoke of burning sesame to relieve fevers, and sesame oil—in which flowers of rosemary have been soaked for 40 days—are prescribed as a cure for earaches.[8]

§ 3 *Literary Sources*

References to sesame in Greek and Latin literature report that sesame grows in the East and is frequently used there as a substitute for olive oil. Sesame seed as it is used in foods is also discussed. The following are some of the more significant and relevant descriptions of sesame.

Aristophanes *Pax* Va
869 Bakery products are made with sesame seeds.

Herodotus Va
1.93 Sesame grows well in Babylonia where it is the only source of oil.

[4] Eckey, *Fats and Oils*, p. 744.
[5] Jamieson, *Fats and Oils*, p. 238.
[6] Eckey, *Fats and Oils*, p. 746.
[7] Jamieson, *Fats and Oils*, p. 237; Weiss, *Castor, Sesame, Safflower*, pp. 506–25.
[8] Hilton-Simpson, *Arab Medicine*, pp. 74, 78; Weiss, *Castor, Sesame, Safflower*, pp. 314–18; Boulos, *Medicinal Plants*, p. 142.

3.48	Sesame and honey cakes are eaten at Corinth.
3.117	Sesame grows in Persia.
Hippocrates	Va
2.518	Mix cumin with white sesame seeds.
Xenophon *An.*	V-IVa
1.2.22	Sesame grows plentifully in Cilicia.
4.4.13	The Greek infantry used sesame oil in place of olive oil while in Armenia.
6.4.6	Sesame grows in Thrace-in-Asia.
Theophrastus *HP*	IV-IIIa
8.2.6	Sesame matures 40 days after flowering.
8.3.3–4	Sesame flowers are leafy; the seed pods grow out where the flowers had been; sesame is a summer crop.
8.5.1–2	Sesame seeds are white and are in seed pods that have various sections.
8.9.3	Sesame exhausts the soil.
Strabo	Ia
16.1.14	The Mesopotamians use sesame oil, but other places do not have it.
16.1.20	The Mesopotamians anoint themselves with sesame.
16.4.26	The Nabataeans use sesame oil in place of olive oil.
Dioscorides	Ip
2.99(121)	Sesame is not good for the stomach and causes bad breath; the Egyptians use oil made from sesame.
Pliny *Nat.*	Ip
6.32(161)	In Arabia oil is obtained from sesame.
15.7(28)	In India oil is made from sesame.
15.7(30)	In Egypt oil is obtained from sesame.
18.22(96)	Sesame is a summer crop that is made into clear oil in India.
18.23(98)	Sesame is soaked in hot water, spread out to dry, rubbed thoroughly, and then soaked in cold water to separate the chaff; finally it must be spread out in the sun to dry.

18.10(49)	Sesame is sown in summer.
18.10(53)	Sesame seeds are enclosed in seed pods.
18.10(58)	Sesame leaves are blood-colored.
18.73(304)	Sesame oil is thick.

Columella — Ip

2.12.5	Per six *sextarii* of sesame planted: after the ground is broken up, three days of plowing and four days of harrowing are required, four days of hoeing and two at a second hoeing, and two days of harvesting are required; the total per six *sextarii* is 15 days.
2.10.18	Sesame is planted early in the season in wet fields but in the fall in dry fields. Sesame grows best in black soil, but will succeed in sandy soil if it is fertile. Sesame is planted in June and July in Cilicia and Syria and is ready for harvest by autumn.
11.2.50	In the overseas provinces, sesame is sown in June or July.
11.2.56	In Cilicia and Pamphylia, sesame is sown in July.
11.2.75	Four or five *sextarii* of sesame per *iugerum* should be planted.

Curtius Rufus — Ip

7.4.23	Alexander's men used sesame oil in place of olive oil.

Galen — IIp

6.547–48	Sesame is combined with honey to make a confection and is sprinkled on breads; sesame oil is very thick.
12.120	Sesame oil is very thick and sticky yet soft.
19.725	Sesame may be substituted for wild vetch.
19.742	Linseed may be substituted for sesame.

Athenaeus — II–IIIp

3.111a	Bread is made with flax and sesame seeds.

Theophrastus and Pliny often use the sesame plant and its seed as a point of comparison for less well known plants that they are describing.[9] Apparently they could presume that their readers were familiar with sesame. References to sesame are also found in many prescriptions.[10]

Given the similarity of the descriptions of sesame in the literary sources with those in modern sources, the identification of σήσαμον (Latin *sesamum*) with sesame is secure: it grows well in many temperate climates, the seed pods grow out where the flowers had been, each seed pod contains numerous white seeds, sesame seeds are used in bakery products, and sesame oil can be used in place of olive oil.

§ 4 *Ancient Civilizations*

In addition to the evidence from Classical literature cited above, frequent notices appear in Babylonian and Assyrian literature,[11] as well as in Jewish literature,[12] of the cultivation and use of sesame oil. Sesame cultivation and sesame oil production in antiquity was especially concentrated in Mesopotamia.[13] An inscription from Telmessus (*OGIS* 55.16) in the time of Euergetes attests the presence of sesame in Asia Minor also.

But for Egypt before the Greek takeover, no evidence of sesame exists;[14] and for Greece and Italy, the Classical authors—though in

[9] E.g.: Theophrastus *HP* 3.13.6; 3.18.13; 6.5.3; Pliny *Nat.* 6.161; 13.118; 22.132; 27.126; Theophrastus alone has more than twenty references to sesame.

[10] Hippocrates 2.518 *et saepe;* Dioscorides 2.99(121); Pliny *Nat.* 22.64(132); 23.49(95); 26.40(67), 68(110); 28.47(168); Celsus 5.15; and Galen 14.142.

[11] For a general discussion see A. Steier, "Sesamon," *RE* II A (1923), cols. 1849–53; in addition see M. G. Raschke, "New Studies in Roman Commerce with the East," *Aufstieg und Niedergang der Römischen Welt: Principat* 9:2 (Berlin: Walter de Gruyter, 1978), p. 906, n. 1014, 1020; p. 924, n. 1112; G. Pettinato, *Die Ölwahrsagung bei den Babyloniern,* Band I: *Einleitung,* Studi Semitici 21, 22 (Rome 1966), p. 43; B. Meissner, *Babylonien und Assyrien,* vol. I (Heidelberg: Carl Winters, 1920), p. 193; J. Hunger, *Becherwahrsagung bei den Babyloniern* in Leipziger semitische Studien I.1 (Leipzig, 1903), p. 13; and Löw, *Pflanzennamen,* pp. 320–22. Cf. Till, *Arzneikunde,* p. 95.

[12] S. Daiches, *Babylonian Oil Magic in the Later Jewish Literature* (London, 1913), p. 25; R. M. Haywood in Frank, *Economic Survey,* vol. 4, p. 133. The Talmud (*Shabbath* 26a, *Nedarim* 53a) reports that in Babylonia the only oil in use is sesame, but that in Egypt and Palestine, sesame oil is very rare.

[13] Levey, *Ancient Mesopotamia,* pp. 85–94, discusses various types and uses of sesame oil; cf. R. J. Forbes, "Chemical, Culinary, and Cosmetic Arts," *A History of Technology* I (Oxford: Clarendon, 1954), p. 288.

[14] Cf. the discussion above (ch. 1 § 7, pp. 30–31) of Keimer's conclusions to the contrary.

some cases describing its cultivation—do not indicate that it is grown there.

The time of the introduction of sesame into Egypt cannot be narrowly defined. The earliest evidence of sesame in Egypt is Ptolemaic, but it must be noted that an early Demotic text from a temple (314–13 B.C., above ch. 1 § 7, p. 31) is an indication of sesame cultivation by the natives very soon after the arrival of the Greeks. It is therefore likely that sesame was brought to Egypt after the Persian conquest of Egypt. The Greeks, having found sesame oil a better alternative to olive oil than castor oil, encouraged the production of sesame in Egypt beginning in the late fourth century B.C.

§ 5 *Occurrences in Papyri*

The references to sesame are listed in appendix E. The list of documents includes numerous accounts, receipts, and letters. Two archives figure prominently: the Zenon collection frequently mentions planting sesame, and the texts from the Serapeum at Memphis are concerned with yearly provision of sesame and castor oil.

Omitting the references to sesame in ***P.Rev.***, the approximate distribution of papyri is as follows:

Zenon archive:	52	mention sesame plants, seed, etc.
	0	mention sesame oil
Papyri of IIIa:	31	mention sesame plants, seed, etc.
	4	mention sesame oil
Papyri of IIa:	15	mention sesame plants, seed, etc.
	1	mentions sesame oil
Serapeum archive:	17	mention sesame oil
	2	mention sesame seed
Papyri of Ia:	2	mention sesame plants, seed, etc.
	0	mention sesame oil

This tabulation quickly reveals the infrequency of references to sesame oil (except in the Serapeum archive) in contrast to the frequent references to sesame plants and seed. As demonstrated above (ch. 1 § 5, pp. 18–24), ἔλαιον, which occurs throughout the papyri, is a general term for oil with a range of meaning broad enough to include sesame oil.

§ 6 *Geographical Distribution*

Of the eighty-nine papyri of the third century (including the Zenon archive), about one-half are explicit as to the provenance of sesame production. Except for the following papyri, the evidence for sesame cultivation is entirely in the Arsinoite nome: ***PSI*** V 518, Apollonius's *dorea* at Memphis; ***P.Hib.*** I 43, sesame delivered to a factory in the Oxyrhynchite nome; ***P.Tebt.*** III pt. II 844, sesame processed at a factory in the Aphroditopolite nome; and ***P.Tebt.*** III pt. II 845, sesame harvested in Phthemphouth.

In the second century sesame is mentioned in four locations other than the Arsinoite nome: ***BGU*** VI 1217, Hermopolis; ***P.Tebt.*** III pt. II 1043, Heracleopolite nome; ***UPZ*** I 20–89, Memphis; and ***O.Petr.*** 44, Thebes. This absence of evidence for sesame production throughout Egypt and concentration of evidence for sesame production in the Arsinoite nome may partially be explained by the provenance of the extant papyri, but it may also suggest the Greek predilection for sesame. ***P.Rev.*** 60–72, however, assigns the majority of sesame cultivation to nomes in the Delta (below, § 8, pp. 64–65).

§ 7 *Agricultural Details*

Sesame was planted from seed as is indicated by the term σπείρω and related forms in ***P.Cair.Zen.*** II 59268 and V 59816, and by the allotments of seed in ***P.Cair.Zen.*** II 59292, V 59850, and ***P.Lond.*** VII 1994 and 1995.

P.Cair.Zen. V 59816 suggests that sesame was planted when it was impossible to plant something else. The text is a letter conveying Apollonius's instructions to the cultivators of his estate; the land was to be cleared, irrigated, and sown: εἰ δὲ μή, ὅσην ἂν μὴ [ ] δ̣ύν[ηι] σπείρειν .[..] σ̣η̣σ̣α̣μ̣οσπορευθῆι. The text does not clarify why sesame was to be planted only when it was infeasible to plant anything else, although sesame's ability to prosper in most soils may be a factor.

P.Cair.Zen. V 59850 records the daily allotments of seed to three Egyptian cultivators:[15] one received 15 choi., one 5, and one 7. These quantities were apparently for planting. The reason for the variation in amounts is not clear; perhaps it was a difference of experience, of terrain, of tools, or of assistants.

Seven papyri from the Arsinoite nome indicate that two crops of sesame per year were possible in the Fayum:[16]

[15] The names are Paapis, Horos, and Pherenouthis.

[16] Schnebel, *Landwirtschaft*, p. 198; a double planting of grain is mentioned in ***P.Cair.Zen.*** II 59155; ibid., pp. 145–60. Cf. Egyptian practice in the 12th century: R. S. Cooper, "Ibn Mammātī's Rules for the Ministries" (Ph.D. diss. University of California, Berkeley, 1973), pp. 118–20, 147–49.

PSI V	499	(22 Feb.)	money for planting sesame
P.Hamb. I	24	(Apr./May)	σήσαμον θερινόν
PSI V	500	(6 July)	preparation for planting
P.Cair.Zen. V	59850	(Aug./Sept)	seed for planting
P.Lille	41	(29 Dec.)	σήσαμον φθινοπωρικόν
	42	(23 Nov.)	σήσαμον φθινοπωρικόν
P.Lond. VII	1991	(Oct.)	σήσαμον φθινοπωρικόν
P.Wisc. II	77	(25 Sept.)	harvesting sesame

Since parts of the Fayum were not subject to the annual flood and since perennial irrigation was possible, there was more flexibility in planting seasons than in the rest of Egypt or other parts of the Mediterranean world. Theophrastus and Pliny give sesame as only a summer crop; however, Columella suggests that it is planted in some places toward the end of summer (above § 3, p. 60). But in the Fayum sesame was clearly a summer and a fall crop, probably planted in June or July and in October or November.

The time of harvest that would be expected based on sesame's growing season is September or October and January or February. Only the former of these is confirmed by the papyri: ***P.Col.*** III 33 (3 Oct.), receipt for sesame; and ***P.Tebt.*** III pt. II 844 (Oct./Nov.), sesame received at a factory.

The manner of harvest is described in two papyri: ***P.Wisc.*** II 77[17] is a record of paying ¾ ob. per day to workers εἰς τοὺς τὸ σήσαμον τίλλοντας; and ***P.Cair.Zen.*** IV 59787 records payment at lines 21 and 70, εἰς τιλμὸν σησάμου, and at lines 57 and 59, εἰς συναγωγὴν σησάμου. As described above, the uneven ripening and bursting of the sesame seed pods requires that the entire plant be harvested; thus this "pulling out" the entire plant is the manner of harvest.[18]

§ 8 *Amounts Cultivated*

P.Rev. 60–72 assigned to all the nomes but one (the Memphite)[19] a number of arou. to be planted in sesame (in contrast to castor which was to be grown in only eight nomes). The extant figures range from 300 to 10,000 arou. per nome for itself and 240 to 8,900

[17] = ***P.Wisc.*** 1 published by W. L. Westermann and A. G. Laird, "A Zenon Papyrus at the University of Wisconsin" in *JEA* IX (1923), pp. 81–90 (= ***SB*** III 6797).

[18] In the Talmud, *Moed Ḳaṭan* 12b mentions "pulling up sesame crops." Flax is also "pulled": ***P.Cair.Zen.*** IV 59782b.32, 121, 161 and ***P.Lond.*** VII 1997.5. Cf. the cutting of sesame plants in Mesopotamia: Levey, *Ancient Mesopotamia*, p. 89.

[19] Memphis and the Memphite are listed separately. I count them as one; I also count the Thebaid as a nome; see appendix A.

arou. per nome for other nomes, for a total of 51,240 arou. plus the number of arou. not extant (see appendix A).

The nomes in the Delta were especially occupied with sesame production. Of the eleven nomes that were to produce sesame for other nomes or for Alexandria, all but one were in the Delta. The Arsinoite was the only nome of Middle Egypt to produce sesame for other than its own use. The highest figure (10,000 arou.) assigned to a nome for itself was the Saite with Naucratis in the Delta. It was also to supply 3,000 art. to Alexandria. In contrast, the whole Thebaid was to plant 3,350 arou. in sesame for its own use but was not responsible to supply any other nomes. Thus, sesame cultivation was principally to be carried out in the Delta.

The Arsinoite was also responsible for a large amount of sesame. It was to plant the highest number of arou.(8,900) for other nomes, but the number of arou. for itself is not extant. Since the Arsinoite was not to receive sesame from other nomes, it would have likely needed at least 1,000 arou. for its own needs, hence its total number of arou. planted in sesame would have exceeded 10,000.

Other third century papyri include an occasional report of land planted in sesame (all from the Arsinoite nome):

P.Lille	5.20–21	90	arou.	260–59
P.Cair.Zen. II	59207.40	5	arou.	255–54
PSI V	522.1, 11	340	arou.	248–47
P.Col. IV	78.24 (= ***Pap.Lugd.Bat.*** XX 38)	356¼	arou.	before 246
P.Petr. III	75.15	261	arou.	234–33
P.Hamb. I	24.8, 12	80	arou.	222
PSI IV	432.3	90	arou.	IIIa
SB VI	9522.2, 4, 6, 12	3	arou.	IIIa

Though these records of arou. planted in sesame do not account for much sesame,[20] there may have been other plots of sesame growing in the same years in the estate of Apollonius and elsewhere in the Arsinoite nome.

P.Petr. III 75 is the most informative papyrus of the above list, for it is an account by a nomarch of the arou. sown in various crops in the Arsinoite nome. Out of a total of 180,014²⁹⁄₃₂ arou. planted in twelve different crops (the papyrus lists thirteen crops, but the figure

[20] ***PSI*** IV 432 is actually a letter to Zenon asking whether to plant sesame on 90 arou.

for κύαμος is lost), only 261 arou. were planted in sesame, in comparison to 134,315½ arou. planted in πυρός. This report, however, is of land planted by 18 January in fall/winter crops; it can therefore be concluded that sesame as a winter crop received little attention in the Arsinoite nome this year or was already harvested by the time of this papyrus. This limited cultivation of fall sesame is confirmed by the absence of reports of harvest in the months of January and February (above § 7, pp. 63–64).

P.Cair.Zen. II 59292, ***P.Lond.*** VII 1994 and 1995 are records of deliveries of seed for the years 33–35 of Philadelphus to Heracleides, the chief farmer of the estate of Apollonius at Philadelphia.[21] Edgar and Skeat, editors of these texts, suggested that this seed was primarily for planting, but might also have been used for other purposes (εἰς τὸν σπόρον occurs several times). Over a three year period Heracleides received a total of 70,590¾ art., principally of wheat and barley; of sesame he received 756¼ art.[22] Sesame accounts for only 1.07% of the total deliveries, and spread over a three-year period provides only 252 art. per year. If ***P.Rev.***'s seed rate of ½ art. per arou. is used (below § 10, pp. 68–70), that means at the most 504 arou. of sesame were planted each year, disallowing other uses of the seed. This amount of land sown in sesame on Apollonius's estate, approximately 5% of the 10,000 arou., is in line with the papyri cited above that record the number of arou. planted.

Records of amounts of sesame harvested are as follows: ***P.Ryl.*** IV 571, 54 art.; ***PSI*** V 518, 331 art.; and ***P.Tebt.*** III pt. II 845, 90½ art.

Some considerable amounts of sesame are reported in storage from Apollonius's estate. In ***P.Lond.*** VII 1991, Cleitarchus keeps a record (for year 34 of Philadelphus) of the amounts of sesame that Heracleides stored at granaries on the estate. The figures for art. of sesame are: 633, 641 7/12, 279½, 200 5/6, and 13. These figures total 1,767 11/12 art. Since for this year 252 art. were given for planting, assuming that the 756¼ art. reported above for three years was evenly divided between the years, 252 art. of seed (if all were planted as seed) yielded 1,767 11/12 art. of produce, for a ratio of 1 to 7 (below § 10, pp. 68–70). Other amounts of sesame in storage are: ***P.Cair.Zen.*** III 59497, 2⅝ art; IV 59717, 1,535 15/24 art.; and IV 59732, 912½ art.

[21] Whether these deliveries of seed were from the government or from granaries on the estate is not clear; ***P.Rev.*** 43.11–18 states that persons holding land *en doreai* are to keep out of their produce a quantity of seed for the next planting.

[22] See the charts published by Skeat with ***P.Lond.*** VII 1994–1995.

Several deliveries of sesame seed are reported in the Zenon archive. The first two of the following papyri are for processing into oil, and the last is a delivery to the crown: ***P.Col.*** III 33, 108 art.; ***P.Cair.Zen.*** II 59223, 100 art., and 300 art.; and IV 59732, 667½ art.

Papyri of the second century that mention sesame are infrequent and generally mention small amounts of sesame:

P.Tebt. III pt. II	832	rents of 1–4½ art.
	1010	rents of 2¼, 10⅙ art.
	1011	receipts of ⅓, ¼ art.
	1012	rents of ¾–3½ art.
UPZ II	186	20 arou. planted
O.Petr.	44	account of 24 art.
O.Wilck.	1520	account of 19½ art.

Two exceptions to these insignificant amounts of sesame are a letter stating that a factory was short 2,088¾ art. of sesame (***P.Tebt.*** III pt. I 728) and an account of payment in rent of 5,754⅓ art. (***BGU*** VI 1217).

§ 9 *Oil Production*

Though ***P.Rev.*** regulates factories, its only information regarding amounts processed is inferred in cols. 60–72. Apparently all the nomes were involved in oil production plus Alexandria (***P.Rev.*** 58.6 = 60.12).

In addition to the references to oil factories in the Oxyrhynchite and Aphroditopolite nomes—in ***P.Hib.*** I 43 and ***P.Tebt.*** III pt. II 844 (above § 6, pp. 62–63)—***P.Cair.Zen.*** II 59223 and III 59412 cite oil factories in the Arsinoite nome. But the only papyrus that gives information about amounts processed is ***P.Tebt.*** III pt. II 844 (256) from the Aphroditopolite nome, a monthly account of amounts received, amounts processed, and oil produced (see table 3). Though this document is fragmentary, it gives much information about oil production. Note in particular that in Mesore, no sesame was received; in Pachon, none was processed; a deduction of 7% is recorded for loss in cleaning; and 2.9 art. of seed yield 1 metr. of oil. Work at this factory was not only at harvest times, for the seed was stored and processed throughout the year. Based on ***P.Rev.*** 46.16, which requires that each mortar at a factory process one art. per day, this factory must have had 6 mortars in order to process 188 art. in one month.

TABLE 3 *Sesame Oil Production in the Aphroditopolite Nome (**P.Tebt.** III pt. II 844)*

Month	Amount on Hand	Amount Received	Amount Deducted	Amount Processed	Oil Produced
Mesore (21 Sept.–20 Oct.)	143 5/8 art.	none		88 art.	
Thoth (26 Oct.–24 Nov.)		500 art.	32 2/3 art.	188 art.	
Phaophi (25 Nov.–24 Dec.)	334 23/24 art.		5 5/24 art.		
Mecheir (25 Mar.–23 Apr.)	83 metr. 10 ch.			112 art.	38 metr. 6 ch.
Phamenoth (24 Apr.–23 May)				120 art.	41 metr. 6 ko.
Pharmouthi (23 May–21 June)				56 art.	19 metr. 3 ch.
Pachon (22 June–21 July)				none	
Payni (22 July–21 Aug.)				80 art.	27 metr. 6 ch.

§ 10 *Yield Possibilities*

On the ratio of sesame seed sown per arou. there are three indications from the papyri. ***P.Rev.*** 41.14–18 states that the *oikonomos* is to allot 4 dr. to purchase seed for each arou. to be sown in sesame. Since column 39 prescribes a value of 8 dr. per art. for sesame, the intended planting ratio was 1/2 art. per arou. Sesame then is an exception to Wilcken's suggestion that one arou. is generally sown with one art.,[23] and this is not unexpected since the size of sesame seed is significantly smaller than most crops. As ***P.Lond.*** VII 2008.33–34 demonstrates, sesame was sometimes sown less generously than 1/2 art. per arou.; in a letter to Zenon, an inspector reports that the sesame crop is very sparse because the cultivator had only sown 4 1/2 choi. per arou. And in ***P.Rainer Cent.*** 47–48(213), 1/8 art. is allotted per arou. to several cleruchies.

Columella 2.10.18 (cf. 2.9.18) gives a rate of 4 or 6 *sextarii* per *iugerum*. The higher of those figures is equivalent to only 0.1 art. per arou. or 4 choi. Modern seed rates, as low as 1 lb. per acre and as high as 10 lb. per acre,[24] are similar to the lower ones of Ptolemaic times. When the 10 lb. per acre figure is converted to art. per arou., using appropriate equivalencies (see ch. 1 § 3, pp. 9–10) and the weight by volume of sesame (1 bu. = 46 lb.[25]), the seed rate is only 0.1 art. or 4 choi. per arou. Two variables may account for some of the variation in rates of seed per land area: the percentage of seed that germinates and the method of planting, whether broadcast or in rows. Nevertheless, the seed rate suggested in ***P.Rev.***, 1/2 art. per arou., is not supported by other information.

The only papyrological evidence for the amount of sesame that might be harvested per arou. comes again from the Zenon archive. If the amounts of sesame seed allotted for three years in ***P.Cair.Zen.*** II 59292, and ***P.Lond.*** VII 1994 and 1995, are being used exclusively for seed, and are equally spread over the three years, and if the total of the amounts of sesame placed in the estate's granaries (***P.Lond.*** VII 1991) for one of those years is the produce of that seed allotted (above § 8, p. 66), then each art. planted yielded 7 art., which equals 3.5 art. from each arou. planted in sesame (1/2 art. per arou.).

Because of some of the uncertain assumptions of the above calculation, a comparison with modern yields for sesame will help evaluate the yield of 3.5 art. per arou. Egypt's average yield of sesame for the years 1934–38 and 1946–48 was 730 kgs. per ha.[26] Using the equivalencies in ch. 1 (§ 3, pp. 9–10) and the weight by volume of sesame (1 bu. = 20.9 kgs.[27]), the yield of sesame in Egypt

[23] ***O.Wilck.*** I, p. 777.

[24] Joshi, *Sesamum,* p. 13.

[25] U.S. Dept. of Agriculture, "Statistics," pp. v–vi.

[26] *FAO 1949 Production Yearbook of Agricultural Statistics* (Washington: Food and Agriculture Organization of the United Nations, 1950), p. 82; *Almanac 1937* (Cairo: Government Press, 1936), p. 424, gives 2.61 ardeb per feddan; more recent yield figures have increased to as high as 1,226 kgs. per ha. for 1975; *FAO 1977 Production Yearbook*, p. 130. Cf. Weiss, *Castor, Sesame, and Safflower,* p. 341.

[27] *U.S. Dept. of Agriculture*, "Statistics," pp. v–vi.

in the mid-twentieth century was equivalent to 6.6 art. per arou.[28] Though this is higher than the 3.5 art. per arou. suggested above, several inequities should be noted. Egypt has a much higher than average yield in modern world production, the world average for 1934–38 and 1946–48 being 373 kgs. per ha., one-half that of Egypt. Sesame yield in India was even less: 237 kgs. per ha. Numerous variables must also be considered when comparing ancient and modern sesame yields: varieties of sesame, differences in climate and soil fertility, artificial fertilization, methods of harvest, etc. Thus, the 3.5 art. per arou. may be fairly representative of sesame yields in Ptolemaic Egypt.

For the yield of sesame oil from sesame seed we have clear figures in ***P.Tebt.*** III pt. II 844; approximately 3 art. of seed yield 1 metr. of oil.

§ 11 *Amounts of Oil*

Outside of the Serapeum archive we have only a few references to sesame oil and only one of those gives an amount: ***P.Tebt.*** III pt. II 997 is a receipt for 2 metr. The frequent references to sesame in the Serapeum archive are requests for 1 metr. of sesame oil for use in the temple. In addition to these references to sesame oil, several occurrences of the phrase ἔλαιον καὶ κίκι (above, ch. 1 § 5, pp. 22–24) also refer to sesame oil.

§ 12 *Seed and Oil Prices*

P.Rev. priced sesame seed at 8 dr. per 30 choi. art. minus a 2 dr. tax. Two other papyri give a price for sesame seed: ***P.Lond.*** VII 1996.42 (250), 6 dr. per art.; and ***PSI*** V 518.6–8 (251–50), 5⅖ dr. per art.

P.Rev. priced sesame oil at 48 dr. per metr. (no other papyri state a price for sesame oil). On the basis that 3 art. yield 1 metr., there is a 100% gross profit possible: 3 art. at 8 dr. each yield oil worth 48 dr. Net profit depended on processing costs, etc., which are unknown.

§ 13 *Uses*

Three papyri reveal that sesame seed was used for purposes other than making oil: ***P.Cair.Zen.*** IV 59562.18 mentions 10 art. of

[28] Wickersham, "Ptolemaic Oilmen," pp. 48–49, assumed that Egypt's current yield ratios are an adequate indicator of yields in Ptolemaic Egypt.

washed sesame that are to be sent along with some other foods for a festival; ***UPZ*** I 89.15 mentions the purchase of some sesame cakes; and ***P.Lille*** 3.53 reports the delivery of 900 art. of sesame in addition to 3,000 art. of barley to a brewery for making beer.[29] No evidence on the use of sesame oil is given in any of the papyri.

§ 14 *Comparisons and Conclusions*

Unlike castor oil, evidence does not exist for sesame oil in Egypt before the fourth century B.C. Sesame cultivation and oil production had been concentrated in but not limited to Mesopotamia. Under Ptolemaic control, however, sesame was actively cultivated in Egypt, at least by the Greeks. The Demotic papyri suggest that the native Egyptians rarely planted sesame.

Sesame was an attractive oil crop, for though it was an annual, it required less work to plant than castor (from seed rather than shoots), its growing season was short enough to allow two crops per year, it grew well on less fertile soil, and it was easily harvested. It also yielded a more attractive oil than castor. Though the occurrences of ἔλαιον that specify sesame oil are infrequent, many of the numerous occasions when ἔλαιον is written in the papyri must be understood as sesame oil.

By Roman times, the picture of oils had changed. The few papyri that mention sesame do not attest to its use as an oil source (see appendix E),[30] though Pliny and Dioscorides— both in the first century A.D.—refer to the use of sesame by the Egyptians.[31] The papyrological evidence clearly suggests that radish seed and olives became the most common oil sources of Roman Egypt.

[29] Rostovtzeff, *Large Estate*, p. 119; cf. sesame beer in Levey, *Ancient Mesopotamia*, p. 92.

[30] Referring only to ***O.Wilck.*** II 763, Johnson includes a discussion of sesame oil as if it were a common oil source in Roman Egypt; Johnson, *Roman Egypt*, p. 506.

[31] On the sources of Pliny: W. Kroll, "C. Plinius Secundus der Altere," *RE* XXI (1951), cols. 299–300, 424–28; of Dioscorides: M. Wellmann, "Dioskurides," *RE* V (1903), cols. 1131–42; Charles Singer, "Greek Biology and Its Relation to the Rise of Modern Biology," *Studies in the History and Method of Science*, vol. II, ed. by Charles Singer (Oxford: Clarendon, 1921), pp. 60–63. On sesame in Egypt in the 12th century, see R. S. Cooper, "Rules for the Ministries," pp. 118–20, 147–49.

IV

Olive
(ἐλαία)

§ 1 *Physical Description*[1]

Olea europaea L. (Fam. Oleaceae) is a small evergeen tree up to 9 m. in height, having many branches with small leathery leaves, dark green on the upper side and light gray on the underside. An olive tree in bloom has small star-shaped flowers that are white with yellow centers. The numerous varieties of olive trees include a wild kind that is small and shrub-like with little value for food or oil.

The fruit of the olive tree is globular or oblong in shape, and is produced in great abundance from each tree (about 115 kgs. in favorable conditions). It is a soft, fleshy fruit with hard shelled pits. The fruit changes from a greenish color to a honey-color to brown or black as it ripens. Olives vary widely in size depending on variety, weather, soil, etc., and may weigh from 1.3 to 20 grams each. Most commonly they are 2–3 cm. in length. The oil content of the flesh is 25–60%. The pit also has oil content, with nearly the same characteristics as the flesh (though in lesser amounts), making it unnecessary to remove the pits before the olives are pressed.

Olive trees require a warm climate without extremes of heat or cold, as is provided by the countries bordering the Mediterranean.

[1] K. G. Bitting, *The Olive* (Chicago: Glass Container Assoc., 1920); Eckey, *Fats and Oils*, pp. 717–27 (note the bibliography, p. 727); Jamieson, *Fats and Oils*, pp. 102–11; Kirschenbauer, *Fats and Oils*, pp. 175–176; California, State Board of Horticulture, *Investigation of the California Olive Industry: Development of the Industry, Unfruitfulness of Varieties, Methods of Culture, Pruning, Processing of Fruit, etc.* (Sacramento, 1900); W. T. Gracey, *Olive Growing in Spain: Description of Varieties Grown, Methods of Cultivation, and the Preparation of Pickled Olives*, Department of Commerce, Special Consular Reports 79 (Washington, 1918); R. Simari and G. B. Martinenghi, *Olivicoltura e oleificio* (Milan, 1950); Trease and Evans, *Pharmacognosy*,

Some rainfall or irrigation is generally necessary in addition to the winter rains, and for best production, cultivation and fertilization is advised. Olives grow well in a diversity of soils, as long as the soil is well drained. Though an olive orchard requires much land to accommodate the widely spaced trees, other crops may be planted between the olive trees. Propagation of olives is most successfully accomplished by cutting out a piece of the trunk of an olive tree or cutting off a branch and planting it. Both ends of the branch are cut and all leaves removed before it is planted. Olive trees can also be started from the tips of branches, from small sprouts that are produced by a mature tree, or by grafting onto an established wild olive tree. Olive seeds that are planted invariably result in a wild tree.

A new olive tree will often produce only small amounts of fruit during its first ten years and will not reach full production for about twenty years. After established, trees will produce for hundreds of years, and in some cases, many centuries. For good harvests, careful pruning is essential because fruit is produced only on branches two years old. Considerable variation in productivity is common for olive trees; weather, lack of pollination, infestations of insects, etc., can adversely affect fruit bearing. Olive production usually follows a two-year cycle: a good crop one year is followed by a poor crop the next.

§ 2 *Modern Production*

The highest worldwide annual production level of the olive industry up to and including 1981 was for the year 1980.[2] Over 11 million m. tons of olives were harvested, and more than 2 million m. tons of olive oil were produced. Italy, Spain, and Greece contributed 65% of the olives harvested, and Turkey, Morocco, and Tunisia contributed another 20%. Quantities produced in other countries include Israel, 45,000 m. tons; U.S.A, 99,000 m. tons; and Egypt, 6,000 m. tons. Though the olive industry in Egypt is comparatively small, the 1960 census[3] reported 216,095 olive trees growing in Egypt of which 189,404 were of productive age.

---

pp. 329–30; Usher, *Dictionary of Plants,* pp. 421–22; Bedevian, *Plant Names,* p. 422; see other sources listed in K. D. White, *A Bibliography of Roman Agriculture,* Bibliographies in Agricultural History 1 (University of Reading, Institute of Agricultural History, 1970), pp. 50–51. Also see below, n. 18.

[2] *FAO 1981 Production Yearbook,* pp. 137–38.

[3] Food and Agriculture Organization of the United Nations, *Report on the 1960 World Census of Agriculture,* vol. I part C (Rome, 1970), p. 210.

OLIVE

The processing of olives for pickling begins with harvest in the months of September, October, or November, depending on the year and on whether green or black olives are desired. Olives for pickling are handpicked and handled carefully to preserve their appearance. Because of the bitter substance present in all olives regardless of maturity, they are inedible and must be treated. By soaking green olives in a lye solution and then curing them in salt brine, olives can be ready to eat in several weeks. A better quality olive can be obtained, however, by soaking in water that is changed at least every other day followed by salt brine. This method requires several months until the bitter substance disappears and is therefore frequently impractical. Ripe olives are usually not treated with lye, but are placed in a solution of salt and various spices for several days until the bitterness disappears. This process can be shortened if the skin of the olive has been punctured.

Olives to be used for oil are picked when the fruit is fully ripe, beginning in December. The best oil comes from fruit that is not overly ripe and still firm. These olives retain considerable moisture and must be artificially dried before crushing. Olives that remain on the trees longer become soft and have little moisture and produce the greatest quantity of oil, though of poorer quality. Olives for oil are harvested in several ways: sometimes they are knocked from the trees by long poles and then gathered and cleaned; often the branches with ripe olives attached are cut from the tree, thus harvesting and pruning at the same time. Without removing the pits, the olives are crushed in a mill and then pressed. The first pressing yields a very pure oil; whereas, subsequent pressings must be refined in settling tanks, centrifuges, and by other processes. In common use early in the twentieth century were stone crushing mills operated by animals and mechanical presses, but these have largely been replaced by more modern equipment.

Olives and their oil are primarily used for edible purposes, being highly prized in the food industry around the world. Olive oil of poor quality is sometimes used in the manufacture of soap and industrial products, but the demand for oil in cooking leads to the refining of even poor quality oil to convert it to an edible product. Some olive oil continues to be used for various medicinal purposes.[4]

[4] Hilton-Simpson, *Arab Medicine,* index, s.v. "olive;" Boulos, *Medicinal Plants,* pp. 137–38; Martindale, *Pharmacopoeia,* p. 1033; W. S. Blackman and R. R. Marett, *The Fellahin of Upper Egypt* (London: George Harrap, 1927), pp. 195, 207.

§ 3 *Literary Sources*

References to olives in Greek and Latin literature are frequent and, in several cases, extensive. Some of the more significant notices and ones relevant to this study are as follows:

Theophrastus *HP*	IV–IIIa
1.9.3	The olive tree is an evergreen.
1.11.4	Because of the fertile soil in Syria, olives are produced in great abundance.
1.14.1	Olives grow on branches that were new the previous year.
2.5.3	Olive branches may be grafted on another olive tree.
2.5.6	Olive trees must be planted at considerable distance apart.
2.7.2	Olive trees need more pruning than most trees.
2.7.3	Fertilization and irrigation are advised for good olive production.
4.2.8–9	Olive trees grow in the Thebaid in Egypt. They are planted at some distance from the Nile, but they are watered by springs.[5] The trees produce very much oil of good quality, though its odor is offensive.
4.3.1.	Olives and olive oil are abundant in Cyrenaica.
5.9.8	Olive wood made into manufactured articles may produce shoots if they get wet.

Cato *Agr.*	III–IIa
3.2	Olives need to be pressed immediately after they are gathered (cf. 64.1–2).
3.5	Two presses are required for an orchard of 100 *iugera* with a mill for each press.
6.1–2	Select the variety of olive to be planted according to the farm's soil and location.
10.1–4	An olive orchard of 240 *iugera* requires 13 workers, three teams of oxen and other animals, presses, mills, vats, three wagons,

[5] Probably the oasis of Khargeh; Rostovtzeff, *SEHHW*, I, p. 355; Schnebel, *Landwirtschaft*, p. 302, is less specific: an oasis of the Libyan desert.

	one harrow, various iron tools, 100 storage jars, etc. (cf. 12.1–13.2).
12–13, 20–22	Elaborate equipment is required for producing olive oil: pitchers, sieves, pulleys, ladles, funnels, wedges, in addition to the mills and presses.
27–28	Plant young olives in large holes at the sowing time, being careful not to disturb the roots.
65.1–2	Green olives produce better oil but ripe olives produce more oil.

Strabo	Ia
11.14.4	Olive trees grow in Armenia (cf. 12.2.1 = Cappadocia; 12.3.30 = Pontus; 16.3.6 = on the coast of the Red Sea).
17.1.35	In Egypt, olive trees are found only in the Arsinoite nome and in some gardens near Alexandria. They are large and productive trees that are capable of yielding a good quality oil, even though its odor is offensive.

Pliny *Nat.*	Ip
13.19(63)	Olives grow in Egypt at an oasis 40 miles from the Nile; it is in the region of Thebes.
15.1(3)	Though getting olive production started used to take many years, it is now possible to harvest olives the first year after trees have been transplanted to the orchard.
15.3(7)	Olive oil loses its flavor if kept for one year.
15.3(11)	Extreme care must be exercised in knocking the fruit out of the olive trees so as not to injure the tree or disturb the buds of the next year's crop.
15.4(15)	The fleshy olives of Egypt are better for food than for oil (cf. 13.19.63).

Columella	Ip
5.8.5	Olive trees cannot endure hot or cold weather, so if in a hot region, plant on the north side of a hill, but if in a cool region, plant on the south side.

5.9.2	The best olive cuttings are the thickness that a man can grasp with his hand; the cutting should be completely submerged in the soil. After three years the new growth should be pruned back; after five years it is ready for transplanting.
5.9.5	Olive trees may also be started from buds that sprout on the tree.
5.9.16	An olive tree that is unproductive will become productive if a branch from a wild olive tree is grafted into it.
11.2.83	October is the time to harvest olives if you want to make green oil.
12.49–51	Olives for eating are prepared in various ways; some must be soaked in hot water, then mixed with fennel, mastic seeds, salt, and must; after three days these olives may be eaten. Others require soaking in vinegar for 40 days.
17.1	Olive trees grow best on hillsides; on the rich soil of the plains the olive produces much foliage but little fruit.

The information about olives is particularly extensive in Columella 5, 12, and 17–20. He discusses in detail ten different varieties of olives, their characteristics, cultivation requirements, and processing particulars. Cato (*Agr.*) describes exactly how to get started in olive farming and in processing olives, with specific measurements, materials, and design of everything needed for working with olives. Pliny (*Nat.* 15) reviews what had been said about olives from Hesiod to Virgil and how the olive industry had changed by his time. He also discusses numerous misunderstandings that had developed about growing olives. Pliny quotes extensively from Cato on how best to grow and process olives.[6]

Olives and olive oil do not receive as much attention among some medical writers as other plants.[7] The medicinal uses of olives were well known and olives were too commonly available to com-

[6] The importance of the olive industry has been carefully noted in such modern treatments as Frank, *Economic Survey* (see Vol. 6 index).

[7] E.g. Dioscorides 1.30(30), 105(138–39), and Galen 11.483; 19.736; but cf. the recipes in Hippocrates 7, "Nature of Women" and 8, "Diseases of Women."

mand as much reverence among the populace for the miracle of healing.

§ 4 *Ancient Civilizations*

Olives were a prominent part of the agriculture of the Graeco-Roman world as demonstrated above; and as Classical sources indicate, olives were grown in many countries surrounding the Mediterranean.[8]

In Egypt, however, before the arrival of the Greeks, there is limited evidence of olives or olive oil.[9] References to olives in Egyptian texts, drawings of olive branches, and the remains of olive twigs preserved in Egyptian tombs are rare until the fourth century B.C.[10] It may be significant then that no Demotic document mentions olives or olive oil. Other sources of oil were more common in pre-Ptolemaic Egypt.

It is also apparent that the Mesopotamian civilizations did not, in general, produce olives or olive oil. They most frequently used sesame oil and have left little evidence of the presence of olive oil.[11] Olive oil, however, was widely used in Palestine[12] and in the northern Mediterranean countries.

§ 5 *Occurrences in Papyri*

P.Rev. makes no reference to olives or olive oil, but from other Ptolemaic papyri we learn that olive trees were being planted and some olives were available. The Ptolemaic references to olive are listed in appendix F.

The distribution of papyri mentioning olive is centered primarily in the Zenon archive. The evidence is limited to the few occurrences of ὀμφάκινος and λευκός (cf. above, ch. 1 § 5, pp. 21–22) and one occurrence of ἀμόργης (below, § 8, pp. 81–82).

[8] A. Deissmann makes much of the prominence of olives in "The World of Paul, the World of the Olive Tree." *Paul: A Study in Social and Religious History,* trans. W. E. Wilson (London: Hodder and Stoughton, 1926), pp. 37–40.

[9] Schnebel's assumption (*Landwirtschaft,* p. 302 n. 4) that olive trees were already established in Pharaonic Egypt is based on sources of the fourth century and after. Cf. Ch. Dubois, "L'olivier et l'huile dans l'ancienne Egypte" (*Revue de philologie,* 2nd ser. 49 [1925]), pp. 60–83.

[10] Lucas, *Egyptian Materials,* pp. 333–35; cf. Dawson, "Studies," p. 65.

[11] Thompson, *Assyrian Botany,* p. 102; Levey, *Ancient Mesopotamia,* p. 87.

[12] Löw, *Pflanzennamen,* pp. 136–38; Frank, *Economic Survey,* vol. 4, pp. 133, 37.

§ 6 *Agricultural Details*

Olives were started from shoots: φυτά, μοσχεύματα, κορμοί.[13] The terms may suggest the different possible parts of the olive tree that could be used, whether a small branch, a thick stem, or a bud.[14]

PSI IV 430 is a fragmentary text that has τὰ γίγαρτα τῆς ἐλαίας. It has been suggested[15] that this is a reference to olive seed. Since the papyrus does not indicate that the context is planting, and since the usual manner of propagation was shoots, that is improbable. It may rather refer to the olive pits remaining after the oil was pressed from the fruit.

No Ptolemaic papyrus reveals how widely spaced the olive trees were in the orchards, but a Roman text, ***BGU*** I 241.28 (IIp) gives a ratio of 18 olive trees per arou.

Three Zenon papyri suggest the times when olives were planted: ***P.Cair.Zen.*** I 59125 (16 Feb.), approval for planting; ***P.Cair.Zen.*** II 59159 (7 Jan.), instructions to plant; and ***P.Cair.Zen.*** II 59244 (11 Feb.), shoots being sent.

Apollonius's instructions for the planting of olive trees on his estate at Philadelphia specify that the olive shoots are to come from his estate at Memphis[16] and are to be either ἀγριέλαιον (***P.Cair.Zen.*** II 59184) or καλλιέλαιον (***P.Cair.Zen.*** I 59125) but not ἡ Αἰγυπτία, because the latter are only good for parks. Olive culture was apparently established at Memphis[17] before the efforts to plant olive trees in the Fayum; furthermore, some selectivity in varieties of olive was possible. The nature of the Egyptian olive, however, is unclear. Theophrastus's complaint about its odor (*HP* 4.2.9) is the only comment extant on any distinguishing characteristics. Whether ἀγριέλαιον was used in grafting to produce more olives as Columella suggests (5.9.16) or whether it was merely a term for a variety of cultivated olives is unclear.

Once a grove of young olive trees was planted, it could not be ignored. ***P.Cair.Zen.*** I 59072 and ***P.Mich.*** I 45 mention the required irrigation and pruning of olive trees. And ***P.Cair.Zen.*** 59734 assigns four yoke of oxen for each arou. (cf. Cato *Agr.* 10.1–4).

[13] ***P.Cair.Zen.*** I 59072; 59125; II 59159; 59184; 59244; III 59431; V 59839; and ***P.Lond.*** VII 1972; cf. ***P.Lond.*** VII 2164.

[14] Skeat on ***P.Lond.*** VII 1972.

[15] Rostovtzeff, *Large Estate*, p. 72.

[16] ***P.Cair.Zen.*** V 59839 reports that olive shoots had been sent to Zenon from Alexandria (cf. Strabo, above § 3).

[17] ***PSI*** VII 862.12 has ἐλαῶν Μεμφιτικῶν.

On harvest time there is one dated papyrus: ***P.Cair.Zen.*** II 59184 (8 October) records Apollonius's instruction to mark the trees before harvest from which shoots are to be taken for new plantings.

§ 7 *Amounts Cultivated*

For the extent to which olive trees were planted, the only data is from Apollonius's estate at Philadelphia:

P.Cair.Zen.	II	59184	Not less than 3,000 shoots
		59244	140 olive shoots
	III	59431	15 olive shoots
	IV	59734	Olive orchard of 88 arou.
	V	59839	470 olive shoots
P.Lond.	VII	1972	As many shoots as possible
		2002	100 dr. to purchase olive shoots

No papyri refer to olives being delivered to factories or being processed. The following papyri are the only records of amounts of olives and they are all for individuals:

P.Cair.Zen.	I	59099	2 petaliai of olives
	III	59501	2 keramia of black olives
PSI	V	535	2 stamnia, four keramia
		553	3 keramia of olives
		594	10 keramia of olives
	VII	858	1 keramion of olives
P.Tebt.	III pt. I	793	2 keramia of olives

From the second and first centuries, no records of amounts of olive cultivation are extant, though there is an occasional mention of olives, of olive groves, and of the planting of olives.

§ 8 *Amounts of Oil*

P.Cair.Zen. 59839.3 mentions 4 keramia of ἀμόργης. Dioscorides 1.102(134) says that this is the sediment (ὑποστάθμη) of pressed olives. Pliny *Nat.* 15.9 says that *amurca* is the bitter fluid of which the olive consists, in addition to the pit, oil, and flesh. Medical writers prescribe ἀμόργης for a variety of problems (e.g. Dioscorides, *op.cit.;* Galen 11.504, 824; Celsus 5.28.16c). What Zenon was to do with this quantity of ἀμόργης is unknown, but it at least provides evidence of the processing of olives.

§ 9 *Uses*

The only thing that is known about the use of olives and their oil is based on occurrences of olives in food lists. Numerous accounts attest to the common place olives held in the diets of the Greeks in Ptolemaic Egypt.

§ 10 *Comparisons and Conclusions*

Olives initially required the greatest amount of labor of any of the oil sources. They had to be planted from shoots and needed yearly attention, and an olive tree did not produce many olives until it was at least ten years old. Olives, however, were a much less common crop than sesame or castor. Apollonius was directing the planting of some olive trees on his estate at Philadelphus, but beyond that there is little evidence of amounts planted. No Demotic evidence exists for olive oil.

The references to olives and their oil in the Roman papyri are numerous. In contrast to castor and sesame, olive oil, which was not in wide use in Ptolemaic Egypt, was by the Roman period more common than any of the oils used in Ptolemaic Egypt.[18]

[18] For additional bibliography on olive oil that came to my attention too late to be included in the discussion above, see: M.-C. Amouretti, *Le pain et l'huile en Grèce antique* (Paris: Les Belles Lettres, 1986); M.-C. Amouretti and G. Comet, *Le livre de l'olivier* (Aix-en-Provence: Edisud, 1985); ibid., *L'olivier in Provence* (Aix-en-Provence: Edisud, 1979); J.-P. Brun, *L'oléiculture antique en Provence* (Paris: Editions du CNRS, 1987); H. Camps Fabrer, *L'olivier et l'huile en Afrique romaine* (Algiers, 1953); ibid., *L'Huile d'olive en Méditerranée. Histoire, anthropologie, économie de l'antiquité à nos jours* (Aix-en-Provence: CNRS, 1985); H. Forbes and L. Foxhall, "The Queen of all Trees. Preliminary Notes on the Archaeology of the Olive," *Expedition* 21.1 (1978) 37–47; M. Heltzer and D. Eitan, eds., *Olive Oil in Antiquity* (Haifa, 1987); D. J. Mattingly, "Oil for Export? A Comparison of Libyan, Spanish and Tunisian Olive Oil Production in the Roman Empire," *Journal of Roman Archaeology* 1.1 (1988) 33–56; ibid., "Megolithic Madness and Measurement, or How Many Olives Could an Olive Press Press?" *Oxford Journal of Archaeology* 7.2 (1988) 177–96; ibid., "Olive Oil Production in Roman Tripolitania" in D. J. Buck and D. J. Mattingly, eds., *Town and Country in Roman Tripolitania* (BAR S274, 1985); ibid., "The Olive Boom. Oil Surpluses, Wealth and Power in Roman Tripolitania," *Libyan Studies* 19 (1988) 21–42; J. J. Rossiter, "Wine and Oil Pressing at Roman Farms in Italy," *Phoenix* 35 (1981) 345–61.

V

Safflower
(κνῆκος)

§ 1 *Physical Description*[1]

Carthamus tinctorius L. (Fam. Compositae) is an erect annual plant 0.5–1.5 m. in height. The safflower plant tends to branch out near its base into several long stems; the stems and leaves are prickly, resembling a thistle. Each of several flowers per stem have 20–100 florets that mature into an equal number of seeds under favorable conditions.

The safflower seeds are generally pear-shaped, somewhat flattened, pale gray to white in color, and in length up to 10 mm. Considerable variation in size and shape is common among different varieties. The seed of the safflower plant does not readily break open, so mechanical harvest does not result in a significant loss of seeds. The oil content of the seed kernel without the hull is about 50%; with the hull, 25–37%.

In India, safflower is sown in October or November. By December or January the plants flower and in two more months are ready for harvest. Normally, the whole plant is pulled from the ground and dried before processing.

Safflower is well adapted to sandy soils and has good tolerance to drought, such that it can be grown in semiarid regions. Its best yield, however, is conditioned on a good supply of soil moisture together with hot and dry weather. Safflower exists today only under cultivation.

[1] Weiss, *Castor, Sesame, and Safflower*, pp. 553–91; Eckey, *Fats and Oils*, pp. 777–78; Jamieson, *Fats and Oils*, pp. 297–98; Vaughan, *Oil Seeds*, p. 40; *Grain and Oil Seeds*, part IV, p. 22; Martindale, *Pharmacopoeia*, p. 1034; Usher, *Dictionary of Plants*, p. 126; Till, *Arzneikunde*, p. 89; Bedevian, *Plant Names*, p. 148.

SAFFLOWER

§ 2 *Modern Production*

The highest worldwide annual production level for safflower seed through 1981 was for the years 1975, 1977, and 1979, at 1 million m. tons.[2] Mexico and India were the two countries producing the largest amounts. None was reported for Egypt, and of the other Mediterranean countries, only small amounts were produced in Israel, Turkey, Portugal, Spain, and Morocco. Two agricultural censuses report that Egypt has had some safflower cultivation in the current century: the total of all Egyptian crops of 1934–35 were planted on 2.7 million ha. of which 775 were in safflower; the 1960 census lists only 123 ha. planted in safflower out of 2.6 million.[3]

Oil is pressed today from safflower seed without removing the hulls and is used primarily in food products, margarine, cooking oils, etc. It is also used in the manufacture of paints and varnishes and is occasionally used in India for illumination and in making soap.

§ 3 *Literary Sources*

There is little doubt that κνῆκος (Latin *cnecus*) is to be identified with safflower. One attempt has been made to the contrary: "Cnecus oil was almost certainly made from the seeds of some composite plant, possibly an artichoke."[4]

Safflower was not a well-known plant in the Greek and Roman world, though it appeared in some medical prescriptions; e.g. Hippocrates 7.364; Pliny *Nat.* 21.107(184); Galen 11.610, 612. The following references are the more significant non-medical notices:

Theophrastus *HP* IV–IIIa

1.13.3	Each petal of safflower is attached to a seed.
6.1.3	Safflower has prickly leaves.
6.4.5	Safflower produces many seeds.

Dioscorides Ip

4.188(190)	Safflower has long prickly leaves; the seeds are white and reddish.

[2] *FAO 1977 Production Yearbook*, p. 132; *FAO 1981 Production Yearbook*, p. 134; cf. Weiss, *Castor, Sesame, and Safflower*, pp. 536–52.

[3] *Almanac 1937* (Cairo: Government Press, 1936), p. 424; *Report on the 1960 World Census of Agriculture* (Rome: Food and Agriculture Organization of the United Nations, 1970), vol. I pt. C, pp. 208–09.

[4] E. P. Wright in Grenfell, *Revenue Laws*, pp. 124–25.

Pliny *Nat.* 21.53(90)	The Egyptians value safflower, which is unknown in Italy, for its oil. Wild safflower has two kinds, one more prickly than the other. There is one kind of cultivated safflower.	Ip
Columella 7.8.1	Safflower seed will curdle milk in order to make cheese.	Ip

§ 4 *Ancient Civilizations*

Other than in Egypt, evidence is lacking for the cultivation of safflower in the ancient world.[5] For Pharaonic Egypt, there are infrequent occurrences of the hieroglyphic word for safflower, but beginning in the XVIIIth dynasty, mummies have been found that were wreathed in the flowers of safflower. No evidence, however, indicates that safflower oil was in use.

§ 5 *Occurrences in Papyri*

The infrequent references to safflower are listed in appendix G. The contexts in which safflower is mentioned in ***P.Rev.*** include seed prices, oil prices, harvest, and deliveries to factories. It does not appear, however, in cols. 60–72 where planting assignments are given per nome.

§ 6 *Amounts Reported*

Despite the few references to safflower, sizable quantities are mentioned:

P.Cair.Zen.	II	59223	105 art. received	III
		59292	80¾ art., 5 choi. allotted	III
P.Tebt.	III pt. II	829	103 and 172 arou. planted	II
	III pt. I	728	726 art. shortage at a factory	II
	I	120	1500 dr. for safflower seed	I

Rent charged in kind for land planted in safflower is $4\frac{45}{48}$ art. per arou., according to ***P.Tebt.*** III pt. II 829.

[5] Keimer, *Gartenpflanzen,* p. 7; and Woenig, *Pflanzen,* pp. 351–53; Weiss, *Castor, Sesame, and Safflower,* pp. 529–35.

Our only evidence for processing safflower for its oil is the report that there is a shortage of 726 art. of safflower at a factory (***P.Tebt.*** III pt. I 728). On the possible yield of oil, a Roman papyrus, ***P.Mich.*** inv. 1347 (*ZPE* 33 [1979], 201-6) ***SB*** XVI 12564, reports the processing of safflower at 1 metr. from 9 art.

In addition, three Ptolemaic references to safflower oil report expenditures for oil: ***P.Tebt.*** III pt. II 997 (early II), 2,160 and 2,500 dr.; ***P.Tebt.*** I 122 (96/63), 80 dr.; ***P.Tebt.*** I 253 (96/63), 60 dr.

§ 7 *Seed and Oil Prices*

P.Rev. 39 and 40 give the price of safflower seed at 1 dr. 2 ob. per art. and the price of safflower oil at 48 dr. per metr. The only other papyri reporting safflower costs are ***P.Tebt.*** I 122 (96/63), an expenditure of 80 dr., apparently for ½ kotyle of safflower oil; and ***P.Tebt.*** III pt. II 997 (early II), expenditures of 2,160 and 2,500 dr., apparently per metr. of safflower oil.[6]

§ 8 *Comparisons and Conclusions*

Of the Ptolemaic oil sources, safflower is mentioned the least frequently and does not occur at all in Demotic papyri. The few references to safflower, however, do suggest that the occasional cultivation of safflower involved considerable amounts.

The numerous Roman papyri that mention safflower (see appendix G) present a mixed picture: approximately one-half of the occurrences are in statements that forbid planting safflower (πλὴν κνήκου).[7] Yet safflower was being grown and processed. ***P.Mich.*** inv. 1347, for example, is a report from a supervisor of an oil press that 195 art. of safflower seed had been received and 22 metr. of safflower oil had been shipped. Like olives, safflower was much more common as a crop and as an oil in Roman than in Ptolemaic Egypt.

[6] On inflation see ch. 2, p. 52, n. 64.

[7] On the prohibition of safflower see D. Hagedorn, "Zum Anbauverbot von ἰσάτις, ὀχομένιον, und κνῆκος," *ZPE* 17 (1975), 85-90.

Appendix A
Nome Lists in ***P.Rev.***

P.Rev. 60–72 assigns sesame and castor production according to nome. Twenty-three nomes are listed plus Memphis, which is listed separately from the Memphite nome; a twenty-fifth entry is the Thebaid, all the district south of the Hermopolite nome. This list of nomes represents, then, Lower and Middle Egypt.

References to other nomes in Ptolemaic Egypt suggest that ***P.Rev.*** 60–72 is an incomplete list. ***P.Rev.*** 31 is a list of nomes in regard to taxation. The two nome-lists within ***P.Rev.***, column 31 and columns 60–72, do not agree in total number and do not list the same nomes (20 nomes are the same in both lists). Several attempts have been made to reconcile the two lists, assuming alternate names for some nomes, and counting Memphis and the Memphite as two separate nomes in one list but not in the other.[1] These attempts have failed to demonstrate that either of the lists is a complete record of the nomes of Ptolemaic Egypt. There is, furthermore, evidence that at least three other nomes were in existence by the time ***P.Rev.*** was written: Xoite, Phthemphouth, and Menelaite.[2]

[1] J. P. Mahaffy in Grenfell, *Revenue Laws*, p. xlv–li; H. Gauthier, *Les nomes d'Égypt* (Cairo: L'Institut français d'archéologie orientale, 1935), pp. 29–47; Grenfell and Hunt on *P.Oxy*. XI 1380.21; A. H. M. Jones, *The Cities of the Eastern Roman Provinces*, 2nd ed. (Oxford: Clarendon Press, 1971), pp. 297–301.

[2] J. D. Thomas, "The Nome Lists in the Papyrus of the Revenue Laws," *Aegyptus* XLVII (1967) 217–21.

Sesame

Nome	Number of ar. planted for itself	Number of art. from other nomes	Amounts for other nomes
Saite with Naucratis	10,000		3,000 art. †
Libya	5,700		[. . .] †
Prosopite	1,800		
Nitriote	300		
Sebennytic	[. . .]		[. . .] †
Mendesian	3,000		2,000 ar.
Busirite	[.],650		1,350 ar.
Athribite	1,500	6,800	
Heliopolite	500	2,000	
Bubastite	1,000	[. . .]	
Arabia	1,040		2,000 ar.
Sethroite	[. . .]		[. . .] ar.
Tanite	1,430		1,570 ar.
Leontopolite	[. . .]		240 ar.
Pharbaethite	[. . .]		[. . .] ar.
Letopolite	480		
Memphis		4,200*	
Hermopolite	[. . .]	12,000	
Oxyrhynchite	1,800		
Heracleopolite	2,000	2,800	
Arsinoite (Lake)	[. . .]		8,900 ar.
Aphroditopolite	630		
Cynopolite	[. . .]	[. . .]	
Memphite		2,400*	
Thebaid	3,350		

* = from the Arsinoite nome † = for Alexandria

Totals of extant figures:	35,180	30,200	16,060 ar. 3,000 art.

Castor

Nome	Number of ar. planted for itself	Number of art. from other nomes	Amounts for other nomes
Saite with Naucratis	11,433⅔		10,666⅓ ar. †
Libya		[. . .]	
Prosopite	2,000		13,600 art. †
Nitriote		4,000	
Sebennytic		10,500	
Mendesian		[.],500	
Busirite	[. . .]		[. . .] ar.
Athribite	3,760		3,000 art.
Heliopolite		3,500	
Bubastite		[. . .]	
Arabia		3,700	
Sethroite		5,460	
Tanite		5,040	
Leontopolite		6,200	
Pharbaethite		[. . .]	
Letopolite	550		1,250 ar.
Memphis		[. . .]*	
Hermopolite		12,900	
Oxyrhynchite		6,6[. .]	
Heracleopolite		9,500	
Arsinoite (Lake)	[. . .]		[. . .] ar.
Aphroditopolite		2,200	
Cynopolite	[. . .]		
Memphite		2,120*	
Thebaid	11,820		9,067 ar. †

* = from the Arsinoite nome † = for Alexandria

Totals of extant figures:	29,563⅔	72,220	20,983⅓ ar. 16,600 art.

Appendix B
Oil
(ἔλαιον)

The occurrences of ἔλαιον (and ἐλαϊκός or ἐλάϊνος when referring to oil) are as follows: first the references in ***P.Rev.*** are given, then arranged by century are the rest of the papyri.

P.Rev.		40.9; 41.12; 45.2,3; 47.2,14; 48.4; 49.18,21; 51.5,12,21,24; 52.1,10,13,22,24,26; 53.2,17,20,27; 54.8,9,10,16; 55.7,15,19; 57.16,18,19 (=59.19,20,21); 58.2 (=60.4); 60.16	
Third Century			
BGU	VI	1300.14	III/II
	X	1990.3	III
P.Alex.		489 (p. 31)	III/II
P.Bon.	II	1(v).10	III
P.Cair.Zen.	I	59008.38	259
		59012.*passim*	259
		59013.13	259
		59015 (context is imported oil)	259
		59025.26	257–56
		59077.2,9	257
	II	59160.2,14	255
		59187.2	255
	III	59321.2	249
		59375.10,11	c. 257
		59440.2,15	III
		59457.5	III
		59507.6	III

	IV	59583.1 (=***Pap.Lugd.Bat.*** XX 25)	257
		59627.8	III
		59654.5	III
		59681.10	258–57
		59682.7	258–57
		59702.*passim*	III
		59706.10,25	III
	V	59809.3,6,7	257
		59836.8	III
P.Col.Zen.	I	4.16	258
		14.1,9 (=***Pap.Lugd.Bat.*** XX 25)	257
	II	66.16	256–55
		75.21	248–46
		90.12,22,24	243–42
		117b.1	III
P.Corn.		1.68,204	256
P.Enteux.		31.4	221
P.Hamb.	II	182.10–12,16	249
P.Hib.	I	41.22	261
		59.7	245
		112.2,39,74	260
		113.12,14	260
		121.*passim*	250
		131	245
	II	248	c. 250
P.Lille	I	3.55,56	241–40
		9.4,8	III
		58.*passim*	III
P.Lond.	VII	2017.14,17	241–40
		2099.3	III
		2140.12,20,25	III
		2162.6	III
P.Mich.	I	2.10	259
		103.24	III
P.Petr.	I	29.7	III
	II	25.*passim*	226
		33a.31	III
		34.*passim*	III
		47a.4; b.4	III
	III	61g.3,7; h.5	
		84.5	III
		137i.*passim*	III

		140a.2; b.7	III
		142.6,14	III
P.Ryl.	II	260.3	III
	IV	556.8,9	III
		562.18,21	III
PSI	IV	372.4,5,7	250
		418.11	III
		428.*passim*	III
		430.2,3	III
		443.6 (=***P.Cair.Zen.*** 59507)	III
	V	528.5	III
		531.5–7	III
		535.*passim*	III
	VI	553.8,9	260
		571.16,17	251
		594.10	III
		613.13	III
		671.8	III
	VII	854.13	258
		858.17,18	III
P.Tebt.	I	136	203
	III pt. I	815. ii r.71; ii v.44; vi.45	III
	pt. II	865.*passim*	III
		885.19,35,59	200
		983.2,5	248
		1078.23,31	III
P.Zen. Pestm.		25.1,10 (=***P.Cair.Zen.*** IV 59583; ***P.Col.Zen.*** 14)	257
SB	III	6319.56,58	III
		7202.18	III
	VIII	9860c.15	III
		9936(35).1,7	III
	XII	10860 iii .2	III
		10863a.4	III
O.Bodl.	I	27.2	III
		28.1	III
		307.11	III
		311v.1	III
O.Stras.		2.2	III
Second Century			
BGU	VI	1422.2	153–52
		1423.2	II

P.Haun.	I	12m.5	II
P.Heid.	III	254	II
P.Iand.	VIII	146 iii 6; iv.13; v.12; vi.1,10; ix.8,16; xi.13	II
P.Lond.	II	219a(r).3	II
P.Ryl.	IV	589.16,76,77	180
P.Stras.	VIII	721.3,12	II
P.Tebt.	I	38.4,10,12,14,20	113
		39.3,9	114
		100.3	117
		112.4,85	112
		116.38,40,44,58	II
		117.15,35,61	105-01
		125	c. 114
		131	100
		212	114-13
		221	112
		234	114
	III pt. II	887.*passim*	II
		889.45,52	II
		890.186	II
		932.5	II
		1082.25,30	II
		1086.11,15	II
	IV	1094.2,[3],5	
SB	I	5627.8,10	III-I
	VI	9425m.5	II
	VIII	9936(39).1,2; (41).20; (44).7	II
UPZ	I	17.8	163
		18.28	163
		19.27,32	163
		20.12,38-40,54,55	163
		21.4	162
		23.6,15-17	162
		24.6,24,25	162
		25.21-24	162
		26.13-16	162
		27.6-9,13-15	162
		28.11	162
		29.3,4,5,10	162
		30.5	162
		31.6	162
		32.11-13,17,18,32	162-61

		33.7,8,13,14	161
		34.4,5,9-11	162-61
		35.10,11,20,23,24	162-61
		36.9,10,17,20,35,36	162-61
		37.10,23	161
		38.4,7,13	161
		39.11	161
		40.9	161
		41.7,16	161-60
		66.4	153
O.Bodl.	I	317.4	II
		354.4	II/I
O.Heid.		254.4	134
O.Mich.		772.3	II/I
		774.1	II/I
		775.4	II/I
O.Theb.		3.3	107
		4.4	107
		5.3	107
First Century			
BGU	VIII	1813.12	I
		1854.8	I
	XIV	2370.65	I
P.Oxy.	XII	1453.17	30-29
P.Tebt.	I	117.15,35,61	99
		121.29,37,74	94/61
		253	96/63
SB	V	7746.22 (=*SEG* viii 531)	57-56
O.Bodl.	I	320v.10	I
O.Camb.		19.7	I

Appendix C
Demotic Papyri

The occurrences of *nḥḥ* (oil), *ꜣḳj* (sesame), and *tgm* (castor) in the Demotic papyri are as follows:

nḥḥ (oil)

I.dem.Nubie		61.12	64p
		417.6	Rom.
O.dem.Leiden		96i.4	Ptol.
		122r.2	Rom.
		140r.4–5	Ptol.
		177ii.3,5	Ptol.
		190.5	Rom.
		195.4	Rom.
		202r.5; v.1	Ptol.
		203v.i.3	Ptol.
		208.5	Ptol.
		223.4	Rom.
		224.4	Rom.
		293.6	Ptol.
		323.6,10	9/10p
		342.5	Ptol.
		471v.2	Ptol.
		494.3	Rom.
O.dem.Medinet Habu		148.1	Rom.
P.dem.Berlin	II	130	Ptol.
P.dem.Cairo	II	31161.2	Ptol.
		31216.9–10,18,19	Ptol.
		31219.3	224–23a
		31225.2,4	Ptol.
		31246.4	Ptol.
		31247.2,11	Ptol.
		31248.2–3	Ptol.
	III	50061a.*passim*	Ptol.

P.dem.Eheverträge		10.2.	315a
		13.2	264a
		14.2	252a
		17.2	230a
		19.4	225a
		20.2	223a
		21.3	221a
		25.3	210a
		27.3	217-190a
		29.3	190a
		35.4	171a
O.dem.Theb.		31.4	188p
		175.7	Rom.
		221.7	208-09p
O.dem.Köln		2.20-21	107a
P.dem.Lille		50	223a
		51	223a
P.London-Leiden			IIIp
	Recto	2.12	
		3.7,9,10,14	
		5.5	
		6.2,8,16,37	
		7.1	
		10.11,12	
		11.22,23	
		12.2-3,5,9,10,13,15,27	
		13.23	
		14.18,20	
		16.23	
		17.22	
		19.30-31	
		20.15,17-19,21-22	
		21.16	
		24.38	
		25.12	
		27.13-14	
		28.5,11	
	Verso	6.1,3	
		8.5	
		9.8	
		11.4	
		13.7	
		16.6	

	20.6	
	33.8	
P.Tebt.Tait	18i.7	Rom.
P.Tor.Amenothes	1.4	171a
P.Vindob.D. 6257	Ph.92	IIp
ꜣḳj (sesame)		
O.dem.Zürich	44.4,11	Ia/Ip
P.dem.Adler	28.7 (=***P.dem.Tempeleide 30***)	Ia
P.dem.Loeb	17.16,21	314–13a
P.dem.Medinet Habu	85.3–4	59p
	93.3–4	6p
	157.6 (=***P.dem.Tempeleide 76***)	Ip
tgm (castor)		
P.dem.Eheverträge	10.2	315a
	17.2	230a
	19.3	225a
	20.2	223a
	21.3	221a
	25.3	210a
	27.3	217–190a
	35.4	171a
O.dem.Theb.	6.2,4	Rom.
	44.5	Rom.
	49.3	Ip
	107.3–4	Rom.
O.dem.Zurich	18.2,4–5	IIa
	19.3–4	Ia
	20.4	Ia
P.dem.Berlin	99	Ptol.
	121	Ptol.
P.dem.Cairo	31219.11	224–23a
P.dem.Ryl.	9.16.18	VIa
P.dem.Tempeleide	42.7	II/Ia
	78.4,7	II/Ia
	79.6	Ia
	88.5	II/Ia
	97.10	Ia/Ip
	117.6–7	Ia
	142.7	118–17a

P.dem.Zen.	1.5–8,10–11, 14, 16,19	256–55a
	3(2).3,5	252–51a
Pap.Lugd.Bat. XX	12.11 (=***P.Col.Zen.*** I 40 and ***P.Cair.Zen.*** IV 59670)	254a
	13.12–13,26,28 (=***P.dem.Zen.*** 3(1).2–3)	252a
P.Tor.Amenothes	1.4	171a

Appendix D
Castor
(κίκι, κροτών)

The occurrences of castor are as follows: first the literary references cited in this monograph are given, next the references in ***P.Rev.***, then arranged by century are the rest of the papyri.

Literary references		
Celsus		5.19.26
Diodorus Siculus		1.34.11
Dioscorides		1.32(38)
		4.161(164)
Galen		12.26
Herodotus		2.94
Hippocrates		7.358
		8.386
Paulus Aegineta		7.3
Pliny	*Nat.*	15.7(25)
		23.41(83)
Strabo		17.2.5
Theophrastus	*HP*	1.101
		(cf. 3.18.7)

References in Papyri

P.Rev.—κροτών	39.3,12,15; 41.11,15,18,21,26; 42.4; 43.14,17,21,23; 44.6; 46.19; 49.17; 53.6,8,10,16; 55.5; 57.6,10,11,13–16 (=59.7,11,13,14,16,18,25); 58.1,4 (=60.4,10); 60.20; 61.7,11,15,22; 62.1,9,14,22; 63.4,8,12,21; 64.10,16; 65.5,10,18; 66.9,14,21;

		67.5,14,19; 68.7,12,17; 69.4,15,20; 70.3,8,18; 71.2,7,15,20; 72.8,14,20,23	
P.Rev.—κίκι		40.10,12,15,16; 41.12; 47.14; 49.18; 51.18,21; 53.8,14,20,21,27; 55.7; 57.18,21 (=59.21,24); 58.2 (=60.5); 60.17	

Other Ptolemaic Papyri

Third century			
BGU	VII	1519.13	III
	XIV	2457.3	III/II
P.Cair.Zen.	II	59176.34	255
		59187.2	255
		59243.14	252
		59251.6	252
		59292.*passim*	250
P.Cair.Zen.	III	59326.3	249
		59412.3,6	III
		59457.6	III
		59494.5	III
		59499.100	III
	IV	59552.3 (=***Pap.Lugd.Bat.*** XX 60)	256
		59629.2,3,5	III
		59635.14,16	III
		59656.5	III
		59670.5 (=***Pap.Lugd.Bat.*** XX 12)	254
		59689.8	257
		59704.*passim*	III
		59705.17,29,38	III
		59706.24,27	III
		59717.17	III
		59760.9	III
		59786.9	III
		59787.*passim*	III
	V	59809.5,8	257
		59810.2,3	257
P.Col.	III	21.2,4	257-56
		37.21	254-50

		40.5 (=***Pap.Lugd.Bat.*** XX 12)	254
	IV	108.3	III
P.Corn.		1.*passim*	256
P.Hib.	I	121.*passim*	251-50
P.Lille	I	3.56	241-40
		58iii.19	III
P.Lond.	VII	1959.2,6	256
		1982.3	252
		1991.*passim*	251
		1994.*passim*	251
		1995.*passim*	251
		2002.36,142	c. 249
		2061.7	III
		2140.2,8	III
		2164.6	III
		2172.25	c. 256
		2188.212	III
P.Mich.	I	72.5	III
P.Petr.	II	25.*passim*	226
		28v.11,32	III
		32(2)b.2,4,8	III
		39a.2,5,9	III
	III	32(g)b.2,4,8	? 241
		43(3).15	239
		75.16	234
		84.7	III
P.Petr.	III	86.6	III
		88.2,5,8	III
		89.4,9	III
		130.1	III
		139a(i).6	III
		140c.2	III

P.Petr.	II	25.*passim*	226
		28v.11,32	III
		32(2)b.2,4,8	III
		39a.2,5,9	III
	III	32(g)b.2,4,8	? 241
		43(3).15	239
		75.16	234
		84.7	III
P.Petr.	III	86.6	III
		88.2,5,8	III
		89.4,9	III

		130.1	III
		139a(i).6	III
		140c.2	III
PSI	IV	349.2,11,12	254–53
		358.5,18,31 (=***Pap.Lugd.Bat.*** XX 13)	252–51
		438.3	III
	V	499.5	257–56
		500.5	257–56
		531.5–8	III
	VI	565.7	255–54
		566.4 (see *BL* III, p. 224)	254–53
	IX	1001.5	256–55
P.Sorb.	I	16.12	257
P.Tebt.	III pt. I	703.102	late III
	pt. II	844.10	256
		845.*passim*	264
		865.2,23,27	III
		885.22,36,58	c. 200
		1056.2	III
P.Wisc.	II	77.3,20,23	254–53
P.Zen.Pestm.		12.5,19 (=***P.Cair.Zen.*** IV 59670; ***P.Col.*** III 40)	254
		13.5,18,31 (=***PSI*** IV 358)	252–51
		60.3,8 (=***P.Cair.Zen.*** IV 59552 and ***SB*** VIII 9667)	256
SB	III	6797.16,19 (=***P.Wisc.*** II 77)	255–54
	IV	7451.3,36	210
	V	7985.3	252–51
		8244.7 (=***Pap.Lugd.Bat.*** XX 56)	252
	VI	9220a.2,11; b.4	254–53
	VIII	9667.3 (see *Archiv* 26 [1978], p. 30 and ***Pap.Lugd.Bat.*** XX 60)	256
	XII	10865.*passim*	III
UPZ	II	158a.*passim*	III
Chrest.Wilck.		304.4,9	III
Second century			
P.Adl.		Gr.7; v.2	106
BGU	VI	1217.27	II
P.Grenf.	I	21.18	126

P.Iand.	VIII	146v.2	II
P.Lond.	VII	2188.212	146
P.Mil.	II	27ii.45	158
P.Ryl.	II	70.1	II
PSI Congr. XVII		22.497r.33	114/78
P.Tebt.	I	5.173	118
		38.12	113
		125	c.114
	III pt.I	728 .7	II
	pt.II	829.47	?180–79
		891.2,13	II
		997.10	early II
	IV	1094.3	114–13
SB	XIV	11338.6	III/II
		11968.16	132–31
UPZ	I	19.32	163
		20.12,38,39,54,55	163
		21.7	162
		23.6,15–17	162
		24.6,24,25	162
		25.21–24	162
		26.13–16	162
		27.6–9,13–15	162
		30.5	162
		32.11–13,17,18,32	162–61
		33.7,8,13,14	161
		34.4,5,9–11	162–61
		35.10,11,20,23,24	162–61
		36.9,10,17,20	162–61
		41.7,16	161–60
		94.*passim*	159–58
		95.1	158
		99.46,50	158
		101.22	156
		104.10,13	II
		119.23,35,39	156
		120.8,13	II
	II	204.3,4	134
O.Bodl.	I	171.5	137
O.Wilck.	II	727.3–6	143
		729.3–5	143
		737.4	140
		741.4	135

		743.5	134
		1608.4	III/II
First Century			
P.Tebt.	II	570	early I
O.Ashm.		3.4	II/I
O.Bodl.	I	361.5	II/I
O.Camb.		19.3	I
O.Theb.		6.2	II/I
O.Wilck.	II	1194v.12	III–I
		1236	III–I
Roman Papyri			
P.Aberd.		29.10	48
P.Erl.		97.4,12	II
P.Leid.	II	(X.4.13), p. 213	III
P.Lips.		97 (32).10	338
P.Lond.	III	856.29	I
P.Soter.		3.6	89–90
P.Stras.	IV	267.10	126
		299r.4	II
O.Bodl.	II	1086.4	I
		1171.4,8	I
		1711.3,5	III
		2053.3	II
		2405.2	II
O.Petr.		195.3	4 B.C.
		196.4	4 B.C.
Stud.Pal.	XXII	177.33	II

Appendix E
Sesame
(σήσαμον)

The occurrences of sesame and its compounds are as follows: first the literary references cited in this monograph are given, next the references in ***P.Rev.***, then arranged by century are the rest of the papyri.

Literary references

Aristophanes	*Pax*	869
Athenaeus		3.111a
Celsus		5.15
Columella		2.10.18
		2.12.5
		11.2.50,56,75
Curtius Rufus		7.4.23
Dioscorides		2.99(121)
Galen		6.547,48
		12.120
		14.142
		19.725,742
Herodotus		1.93
		3.48,117
Hippocrates		2.518
Pliny	*Nat.*	6.32(161)
		13.118
		15.7(28,30)
		18.10(49,53,58); 22(96); 23(98); 73(304)
		22.64(132)
		23.49(95)
		26.40(67); 68(110)
		27.126
		28.47(168)

Strabo		16.1.14,20
		16.4.26
Theophrastus	*HP*	3.13.6
		3.18.13
		6.5.3
		8.2.6
		8.3.3,4
		8.5.1,2
		8.9.3
Xenophon	*An.*	1.2.22
		4.4.13
		6.4.6

References in Papyri

P.Rev.	39.11,15,16,17; 40.10,15; 41.10,15,17,21,26; 42.3; 43.13,17,19,20,23; 44.6; 46.16,18; 49.16,18; 51.13; 53.5,7,10,14,16,27; 55.4,7,15; 57.6,7,10,11,12,14,16,18 (=59.7,11,12,15,16,17,18,19,20, 21); 58.1,2,4 (=60.4,5,9); 60.19,24; 61.2,14,21; 62.4,17; 63.7,13,15,19; 64.4,8,21; 65.3,14; 66.4,17; 67.9; 68.2,16; 69.3,9,13; 70.2,12,16; 71.6,12,14; 72.2,6,13,19

Other Ptolemaic Papyri

Third Century

P.Bon.		11 1 (v). 8,14	III
P.Cair.Zen.	II	59207.40	255–54
		59223.1,6,7	254
		59225.8	253
		59232.7	253
		59251.6	252
		59268.37	252–51
		59292.*passim*	250
	III	59314.4	250
		59315.5	250
		59326.23,165	249
		59412.3	III

		59494.6,7,15	III
		59497.9	III
		59502.10	III
	IV	59562.18	253
		59596.13	III
		59608.4	III
		59656.6	III
		59717.6,13,15	III
	IV	59730 (see ***P.Lond.*** VII 1991)	
		59732.2	III
		59787.21,57,60,71	III
	V	59816.6	257
		59850.*passim*	III
P.Col.	III	33.4	255
		53.3	250
	IV	69.1	257-49
		78.24 (=***Pap.Lugd.Bat.*** XX 38)	before 246
		116c.3	III
P.Enteux.		55.9	221
		58.19,21	221
P.Hamb.	I	24.8,12	222
P.Hib.	I	43.3,5,12	261
		119.18	c. 260
	II	245.2,6	c. 250-40
		247.19	c. 250-40
P.Lille	I	3.53	241
		5.20,21	260
		9.13,20,25	285-46
		26.1	III
		41.4,14	250
		42.4	250
P.Lond.	VII	1991.*passim*	251
		1992.22	251
		1994.*passim*	251
		1995.*passim*	251
		1996.42	250
		2002.131	249
		2003.2	249
		2008.22	247
		2151.3	III
		2172.27-29,31	256
P.Mich.	I	30d.2,13	256
		40.6	254

		43.3,6	253
		53.11	249
		96.3,5	III
P.Petr.	III	43(3).15	239
		75.15	234
		80a (ii).14,20	III
		84.5	III
		95 (ii).12	III
		109d.3	III
P.Rainer.Cent.		47 and 48.*passim*	213
P.Ryl.	IV	571.6,7,10	III
PSI	IV	351.17	254–53
		370.9	250–49
		372.2	250–49
		375.3	250–49
		432.3	III
		438.3	III
	V	499.5	257–56
		500.4	257–56
		502.28	257–56
		518.6	251–50
		522.1,11	248–47
	VII	863.19	III
P.Sorb.		32.10	247
P.Tebt.	III pt. I	701.246,77,80,87	235
		815 (3)r.10; (5).25; (7).6,44	228–21
	pt. II	844.*passim*	256
		845.*passim*	264
		946.13	late III
		983.2,5	248
P.Wisc.	II	77.7,24 (=***SB*** III 6797)	254–53
P.Zen.Pestm.		38.35 (=***P.Col.*** 78)	before 246
SB	I	4369	III
	III	6797.3,20 (=***P.Wisc.*** II 77)	255–54
	VI	9522.2,4,6,12	III
	XII	10865.*passim*	III
O.Theb.		7.4	late III
Second Century			
BGU	VI	1217.26,34,38	II
P.Tebt.	III pt. I	713.2	late II
		728.7	II

	pt. II	832.*passim*	II
		886.73	c. 182
		893.1	II
		913.2	139
		997.6,9	early II
		1010.2,3	180–45
		1011.5,10,13	145–16
		1012.12,16,18,20	145–16
		1043.42,45,48,54	c. 170
		1057.4,5	145–16
UPZ	I	20.12,38,39	163
		21.7	162
		23.6	162
		25.21–24	162
		26.13–16	162
		27.6–9	162
		29.4,5,10	162
		30.5	162
		31.6	162
		32.17,18	162–61
		33.7,8,13,14	161
		36.35,36	162–61
		37.10	161
		38.4,7,13	161
		39.11	161
		40.9	161
		41.7	161–60
		62.20	161
		89.15	160–59
	II	186.1,2	133
O.Petr.		44.3,5	120
O.Wilck.		1520.3	136

First Century

BGU	XIV	2439.55,58,67	I
		2441.101	I

Roman Papyri

P.Lond.	II	222.88	IV
P.Oxy.	XXXVI	2797.12	III/IV
	XLII	3066.3	III
O.Bodl.	II	1714.2,6	II

O.Giss.		Inv. Nr. 22 (*JJP* 16/17 [1971] 100) = ***SB*** XII 10896	110
O.Stras.		475.1	IV/V
		606.2	II
O.Wilck.	II	763	9 B.C.
Stud.Pal.	X	123.1	VIII

Appendix F
Olive
(ἐλαία, ἐλάα)

The references to olives (including ἐλαϊκός or ἐλάϊνος when referring to olives) are as follows: first the literary references cited in this monograph are given, then arranged by century are the papyri.

Literary references

Aristotle	*HA*	520.18
Cato	*Agr.*	3.2,5
		6.1,2
		10.1–4
		12,13
		18–22
		27,28
		64–69
Celsus		5.28.16c
Columella		5.8,9
		11.2.83
		12.49–51,52–54
		17–20
Dioscorides		1.30(29,30,32); 32(38); 33(39); 102(134); 105(138,39)
Galen		11.483,504,824
		13.953
		19.736
Hippocrates		8.306,308,376,386
Pliny		13.19(63)
		15
Strabo		11.14.4
		17.1.35
Theophrastus	*HP*	1.9.3
		1.11.4

		1.14.1	
		2.5.3,6	
		2.7.2,3	
		4.2.8,9	
		4.3.1	
		5.9.8	

References in Papyri
Third Century

P.Cair.Zen.	I	59072.4	257
		59099.3	257
		59125.3,6,7	256
	II	59157.2	256
		59159.2	255
		59184.2,5,9 (see *BL* IV, p. 27)	255
		59244.10	252
P.Cair.Zen.	III	59326.205	c. 249
		59431.4	III
		59501.8	III
	IV	59544.2	257
		59690.5	258
		59734.1	III
		59788.18,27	III
	V	59838.3	III
		59839.3	III
P.Hal.		1.98	III
P.Hib.	I	49.8,12	c. 257
P.Lond.	VII	1972.2	254
		2002.128	249
		2164.8	III
P.Mich.	I	2.10	259
		45.26	252–51
		103.24	III
	III	173.8,23	late III
PSI	IV	428.*passim*	III
		430.2,3	III
	V	535.*passim*	III
	VI	553.8,9	260–59
		594.11	III
		671.8	III
	VII	858.17,18	III
		862.12	III
P.Tebt.	III pt. I	815.*passim*	228–21

P.Yale	I	32.r.8,12 (=***P.Hib.*** 49)	257
SB	I	4369b.8	III
	VI	9091.2	247
UPZ	II	158a.24	III
Second Century			
P.Lond.	VII	2188.101	148
P.Mich.	III	182.16	182
PSI	XIII	1313.10	II
P.Tebt.	I	81.30,31	late II
		82.*passim*	115
		87.43	late II
	III pt. I	793vi.5,7	183
	pt. II	885.19,35,59	c. 200
SB	III	7188.15	154
	VIII	9936.*passim*	III/II
	XIV	11608.4	II
First Century			
O.Edfou		364.2	I

Appendix G
Safflower
(κνῆκος)

The occurrences of safflower and its compounds are as follows: first the literary references cited in this monograph are given, next the references in ***P.Rev.***, then arranged by century are the rest of the papyri.

Literary references		
Columella		7.8.1
Dioscorides		4.188(190)
Galen		11.610,612
Hippocrates		7.364
Pliny	*Nat.*	21.53(90); 107(184)
Theophrastus	*HP*	1.13.3
		6.1.3
		6.4.5

References in Papyri	
P.Rev.	39.5,12; 40.10; 42.4; 43.18; 44.6; 46.17,20; 49.17,18; 53.10,15,17,22; 55.5,8

Other Ptolemaic Papyri

Third Century			
P.Cair.Zen.	II	59223.4	254
		59292.*passim*	250
P.Lond.	VII	1994.*passim*	251
		1995.*passim*	251
SB	I	4369.*passim*	III

Second Century

P.Tebt.	III pt. I	728.7	II
	pt. II	829.*passim*	? 180–79
		997.8,11	early II
O.Wilck.	II	730.3,4	143
		1353.4–6	144

First Century

BGU	XIV	2441.156,248	I
P.Ryl.	II	69.10,15	34
P.Tebt.	I	120.47	98/64
		122.11	96/63
		253	96/63
O.Camb.		19.4	I

Roman Papyri

The following list of Roman papyri marks with an asterisk those that exclude safflower (πλὴν κνήκου):

BGU	II	661.17 see *BL* I, p. 439	140*
	XI	2124.3	II*
	XV	2484.4	II*
CPR	I	31.24 see *BL* I, p. 117	II*
P.Aberd.		181.11	I
P.Amh.	II	91.15	159*
P.Apoll.		93b.15	VIII
		95a.10	VIII
P.Cairo		3049/36ii.27	144*
P.Cair.Masp.	III	67325i(r).10; ii(r).3; (v).3	c. 585
P.Coll.Youtie	I	27.17	165
P.Fam.Teb.		47.16	195*
P.Fay.		101ii.*passim*	18 B.C.
P.IFAO	I	10.14	I*
P.Hamb.	I	64.19	104*
P.Haun.	II	17.r.29	II
P.Mert.	III	107.19	II*
P.Mich.		inv. 1347 (*ZPE* 33 [1979], pp. 201–206)	146
P.Mil.Vogl.	II	104.21	127–28*
	VI	267.16	128*
P.Oxy.	XVI	2058.33	VI
P.Phil.		15.15	153–54*

P.Ross.Georg.	II	29.3	II
P.Ryl.	II	145.17	I
P.Soter.		3.25	89–90*
		4.13	87*
P.Stras.	VI	535.14	II*
P.Tebt.	II	375.15	140*
SB	XIV	11718.1	II*
O.Stras.		313.3,5,6	14
		327.4,6,7	34
O.Fay.		16.2	c. 1

BIBLIOGRAPHY

Notes: 1) The shortened titles used in the footnotes are indicated in this bibliography by bold faced type.
2) This list of sources excludes the publications of papyri (see pp. viii–ix).
3) This list of sources excludes the following reference works: *RE*, *OCD*, *DarSag*, *LSJ*, *OLD*, *TGL*, *EB* (see list of abbreviations).

Almanac 1937. Cairo: Government Press, 1936.

Alpin, P. *Plantes d'Egypte*. L' Institut français d'archéologie orientale, 1980 reprint of 1640 ed.

Andréadès, A. "Des droits de douane prélevés par les Lagides sur le commerce extérieur." *Mélanges Gustave Glotz*. Paris: Les Presses universitaires, 1932.

Baedeker, K. ***Egypt and the Sudan:*** *Handbook for Travelers*. 8th ed. rev. Leipzig: Karl Baedeker, 1929.

Bagnall, W. S. "The Archive of Laches: Prosperous Farmers of the Fayum in the Second Century." Ph.D. diss., Duke University, 1974.

Bahl, J. C. *The Oilseed Trade of India*. 4th ed. Bombay: New Book, 1938.

Bedevian, A. K. *Illustrated Polyglottic Dictionary of* ***Plant Names.*** Cairo: Argus & Papazian, 1936.

Bingen, J. "**Les colonnes 60–72** du P.Revenue Laws et l'aspect fiscal du monopole des huiles." *CE* 41 (1946) 127–48.

______. "Economie grecque et société égyptienne au IIIe siécle." *Das ptolemäische Ägypten*, ed. by H. Maehler and V. M. Strocka. Mainz am Rhein: Philipp von Zabern, 1978.

______. "Papyrologica: P.Oxy. 2191, P.Heid. 244, P.Petrie II 39 (a)." *CE* XXXIX (1964) 167–73.

______. ***Le Papyrus Revenue Laws***—*Tradition grecque et adaptation hellénistique*. Rheinisch-Westfälische Akademie der Wissenschaften, Vorträge, G 231. Opladen: Westdeutscher, 1978.

Bitting, K. *The Olive*. Chicago: Glass Container Assoc., 1920.

Blackman, A. M., and Fairman, H. W. "The Myth of Horus at Edfu–II." *JEA* 29 (1943) 2–36.

Blackman, W., and Marett, R. R. *The Fellahin of Upper Egypt: Their Religious and Industrial Life Today with Special References to Survivals from Ancient Times*. London: George Harrap, 1927.

Bonneau, D. *La crue du Nil, divinité égyptienne à travers mille ans d'histoire*. Paris: Librairie C. Klincksieck, 1964.

______. *Le fisc et le Nil, incidences des irrégularités de la crue du Nil sur la fiscalité foncière dans l'Égypte grecque et romaine*. Publications de l'institut de droit romain de l'université de Paris, n.s. 2. Paris: Editions Cujas, 1972.

______. "Le Nil à l'epoque ptol." *Travaux de la maison de l'Orient, II: L'homme et l'eau en Mediterranee et au Proche Orient*. Lyon, 1981.

______. "Proposition de renouvellement de bail d'une huilerie." *Scritti in onore di Orsolina Montevecchi*. Bologna: CLUEB, 1981.

Boulos, L. *Medicinal Plants of North Africa*. Algonac, MI: Reference Publications, 1983.

Brehaut, E. *Cato the Censor: On Farming*. New York: Columbia, 1933.

Brown, F., Driver, S. R., and Briggs, C. A. *A Hebrew and English Lexicon of the Old Testament based on the Lexicon of W. Gesenius*. Oxford: Clarendon, 1907.

Brunton, G., and Caton-Thompson, G. *The Badarian Civilization*. London: British School of Archaeology in Egypt, 1928.

Burtis, E. L. "The Fats and Oils Economy of India." Foreign Agricultural Service, U.S. Department of Agriculture. July, 1960.

Burton, A. *Diodorus Siculus, Book I, A Commentary*. Études préliminaires aux religions Orientales dans l'empire Romain. Vol. 29. Leiden: Brill, 1972.

California, State Board of Horticulture, *Investigation of the California Olive Industry: Development of the Industry, Unfruitfulness of Varieties, Methods of Culture, Pruning, Processing of Fruit, etc.* Sacramento, 1900.

Candolle, Alph. de. *Origine des plantes cultivées*. Paris: Félix Alcan, 1896.

Canvane, P. "Dissertation on the Oleum Palmae Christi sive Oleum Ricini (Commonly called Castor Oil) . . ." 2nd ed. London: Robinson & Roberts, 1769.

Caton-Thompson, G., and Gardner, E. W. *The Desert Fayum*. Royal Anthropological Institute, 1934.

Charpentier, G. *Recueil de matériaux épigraphiques relatifs a la botanique de l'Égypte antique*. Paris: Trismégiste, 1981.

Cooper, R. S. "Ibn Mammātī's **Rules for the Ministries:** Translation with Commentary of the *Qawānin Al-Dawāwīn.*" Ph.D. diss., University of California: Berkeley, 1973.

Crawford, D. J. ***Kerkeosiris:** An Egyptian Village in the Ptolemaic Period.* Cambridge: University, 1971.

Crum, W. E. *A Coptic Dictionary.* Oxford: Clarendon, 1939.

Daiches, S. *Babylonian Oil Magic in the Later Jewish Literature.* London, 1913.

David, M., and van Groningen, B. A. *Papyrological Primer.* Leiden: E. J. Brill, 1940.

Dawson, W. R. "**Studies** in Medical History: (a) The Origin of the Herbal, (b) Castor-oil in Antiquity." *Aegyptus* 10 (1929) 47–72.

Deissmann, A. "The World of Paul, the World of the Olive Tree." *Paul: A Study in Social and Religious History.* trans. W. E. Wilson (London: Hodder and Stoughton, 1926).

Delitzsch, F. *The Hebrew Language Viewed in the Light of Assyrian Research.* London: Williams and Norgate, 1883.

Dittenberger, W. *Orientis Graeci Inscriptiones Selectae,* I, II. Leipzig: S. Hirzel, 1903–05.

Drachmann, A. G. "Ancient Oil Mills and Presses." *Det Kgl. Danske Videnskabernes Selskab., Archaeologisk-kunsthistoriske Meddelelser,* I.1 Copenhagen: Levin and Munksgaard, 1932.

Dubois, C. "L'olivier et l'huile dans l'ancienne Égypte." *RPh* 2nd ser. 49 (1925) 60–83.

Duncan-Jones, R. P. "The Choenix, the Artaba and the Modius." *ZPE* 21 (1976) 43–52.

———. *The Economy of the Roman Empire: Quantitative Studies.* 2nd ed. Cambridge: Cambridge University, 1982.

———. "Variation in Egyptian Grain Measure." *Chiron* 9 (1979) 347–75.

Dupont-Sommer, A. "Note sur le mot *TQM* dans les ostraca araméens d' Éléphantine." *Semitica* 14 (1964) 71–72.

Eckey, E. W. *Vegetable* ***Fats and Oils.*** American Chemical Society Monograph Series No. 123. New York: Reinhold, 1954.

Edgar, C. C. "A New Letter of Apollonios the Dioketes." *Archiv für Papyrusforschung* 11 (1935) 218–19.

———. "Selected Papyri from the Archives of Zenon." *Annales du service des antiquités de l'Égypte* XXIII (1923) 86–95.

Emory, W., and Wolf, J. S. "**A Study of Practices** Affecting the Use of Major Vegetable Oils for Refining and Processing." St. Louis: Washington University, 1960.

Erichsen, E. *Demotisches Glossar.* Kopenhagen: Munksgaard, 1954.

Erman, A., and Grapow, H. *Ägyptisches Handwörterbuch.* Berlin: Reuther & Reichard, 1921.

______. *Wörterbuch der ägyptischen Sprache*. Leipzig: Hinrichs, 1926.

FAO 1949 ***Production Yearbook.*** FAO Statistics Series. Rome: Food and Agriculture Organization of the United Nations, 1950.

FAO 1966 ***Production Yearbook.*** FAO Statistics Series. Rome: Food and Agriculture Organization of the United Nations, 1967.

FAO 1977 ***Production Yearbook.*** FAO Statistics Series No. 15. Rome: Food and Agriculture Organization of the United Nations, 1978.

FAO 1981 ***Production Yearbook.*** FAO Statistics Series No. 40. Rome: Food and Agriculture Organization of the United Nations, 1982.

Farzat, H. "Encore sur le mot *TQM* dans les documents Araméens d'Éléphantine." *Semitica* 17 (1967) 79.

Faulkner, R. O. *A Concise Dictionary of Middle Egyptian*. Oxford, 1962.

Finley, M. I. *The Ancient Economy*. Berkeley: University of California, 1973.

Food and Agriculture Organization of the United Nations. *Report on the 1960 World Census of Agriculture*. Rome, 1970.

Forbes, R. J. "Chemical, Culinary, and Cosmetic Arts." *A History of Technology*, I. Oxford: Clarendon, 1954.

Frank, T., ed. *An* ***Economic Survey*** *of Ancient Rome*, 6 vols. Baltimore: John Hopkins, 1933–40.

Gara, A. *Prosdiagraphomena e circolazione monetaria*. Testi e documenti per lo studio dell'antichita' LVI. Milan: Cisalpino-Goliardica, 1976.

Gardiner, A. *The Admonitions of an Egyptian Sage from a Hieratic Papyrus in Leiden*. Hildesheim: Georg Olms, 1969.

George, P. S., Srivastava, U. K., and Desai, B. M. *The Oilseeds Economy of India: An Analysis of Past Supply and Projections for 1985*. Delhi: MacMillan of India, 1978.

Germer, R. *Flora des pharaonischen Ägypten*. Deutsches Archäologisches Institut: Abteilung Kairo: Sonderschrift 14. Mainz am Rhein: Philipp von Zabern, 1985.

Gracey, W. T. *Olive Growing in Spain: Description of Varieties Grown, Methods of Cultivation, and the Preparation of Pickled Olives*. Department of Commerce, Special Consular Reports 79. Washington, 1918.

Grain and Oilseeds *Merchants' Association*, Diamond Jubilee. Bombay: Grainseeds House, 1960 (for private circulation only).

Grant, J. "A Note on the Materials of Ancient Textiles and Baskets." *From Early Times to the Fall of Ancient Empires*. Vol. I of *A History of Technology*. Oxford: Clarendon, 1954.

Grapow, H., and von Deines, H. ***Wörterbuch*** *der ägyptischen Drogennamen*. Grundriss der Medizin der alten Ägypter 6. Berlin: Akademie, 1959.

Grelot, P. "L'huile de ricin à Éléphantine." *Semitica* 14 (1964) 64-70.

Grenfell, B. P., and Mahaffy, J. P. *The Revenue Laws of Ptolemy Philadelphus.* Oxford: Clarendon, 1896.

Hagedorn, D. "Zum Anbauverbot von ἰσάτις, ὀχομένιον, und κνῆκος," *ZPE* 17 (1975) 85-90.

Halleux, R. *Les Alchimistes Grecs.* Tome 1 (Budé). Paris: 1981.

Hayes, W. C. "Inscriptions from the Palace of Amenhotep III." *JNES* X (1951) 82-111.

———. *The Scepter of Egypt: A Background for the Study of the Egyptian Antiquities in the Metropolitan Museum of Art.* Part II: *The Hyksos Period and the New Kingdom (1675-1080 B.C.).* Cambridge: Harvard University, 1959.

Helck, W., and Westendorf, W. *Lexikon der Ägyptologie.* Wiesbaden: Harrassowitz, 1972-.

Hilton-Simpson, W. *Arab Medicine and Surgery: A Study of the Healing Art in Algeria.* London: Oxford University, 1922.

Hönigsberg, P. "Römischen Ölmühlen mahlen noch in Oberägypten." *MDAI* Kairo 18 (1962).

Hunger, J. *Becherwahrsagung bei den Babyloniern,* in Leipziger semitische Studien I.1. Leipzig, 1903.

Jahn, J. "Zum Rauminhalt von Artabe und Modis castrensis." *ZPE* 38 (1980) 223-28.

Jamieson, G. S. *Vegetable Fats and Oils: Their Chemistry, Production, and Utilization for Edible, Medicinal, and Technical Purposes.* 2nd ed. American Chemical Society Monograph Series No. 58. New York: Reinhold, 1943.

Janson, H. *Castor Oil Production and Processing. New York: United Nations, 1974.*

Janssen, J. J. Commodity Prices from the Ramessid Period: An Economic Study of the Village of Necropolis Workmen at Thebes. Leiden: Brill, 1975.

Johnson, A. C. *Roman Egypt: To the Reign of Diocletian.* Vol. II of *An Economic Survey of Ancient Rome,* ed. by T. Frank. Baltimore: Johns Hopkins, 1936.

Joshi, A. B. *Sesamum.* Bombay: Indian Central Oilseeds Committee, 1961.

Kamal, H. *A Dictionary of Pharaonic Medicine.* Cairo: National Publication House, 1967.

Keimer, L. "Bemerkungen und Lesefrüchte zur altägyptischen Naturgeschichte." *Kêmi* 2 (1929) 84-106.

———. *Die Gartenpflanzen im alten Ägypten.* Berlin: Ägyptologische Studien, 1924.

Kirschenbauer, H. G. *Fats and Oils: An Outline of Their Chemistry and Technology.* 2nd ed. New York: Reinhold, 1960.

Kosack, W. *Historisches Kartenwerk Ägyptens*. Bonn, 1971.

Levey, M. *Chemistry and Chemical Technology in Ancient Mesopotamia*. Amsterdam: Elsevier, 1959.

Loret, V. *La flore Pharaonique*. Paris: Ernest Leroux, 1892.

Löw, I. *Aramaeische Pflanzennamen*. Leipzig: Wilhelm Engelmann, 1881.

Lucas, A. *Ancient Egyptian Materials and Industries*. 4th ed. rev. by J. R. Harris. London: Edward Arnold, 1962.

Martindale, W. *The Extra Pharmacopoeia*. 27th ed. London: Pharmaceutical Press, 1977.

Maspero, H. *Les finances de l'Egypte sous les Lagides*. Paris, 1905.

Mattil, K. F. "Review and Comparative Analysis of Oilseed Raw Materials and Processes Suitable for the Production of Protein Products for Human Consumption." New York: United Nations, 1974.

Mayser, E. *Grammatik der griechischen Papyri aus der Ptolemäerzeit*. Leipzig und Berlin, 1906–38.

Meissner, B. *Babylonien und Assyrien*. Vol. I (Heidelberg: Carl Winters, 1920).

Montevecchi, O. *La Papirologia*. Torino: Società Editrice Internazionale, 1973.

Moritz, L. A. *Grain Mills and Flour in Classical Antiquity*. Oxford: Clarendon, 1958.

Oil Seeds in India: 1956–57 to 1960–61. New Delhi: Ministry of Food and Agriculture, 1962.

Oleaginous Products and Vegetable Oils: Production and Trade. Rome: International Institute of Agriculture, Bureau of Statistics, 1923.

Patterson, R. "Spinning and Weaving." *The Mediterranean Civilizations and the Middle Ages*. Vol. II of *A History of Technology*. Oxford: Clarendon, 1956.

Patton, W. R., and Myres, J. L. "On Some Carian and Hellenic Oil Presses." *JHS* 18 (1898) 209–17.

Pearl, O. M. "Varia Papyrologia." *TAPA* 71 (1940) 372–90.

Petrie, W. M. F. *Hawara, Biahmu, and Arsinoë*. London: Field and Tuer, 1889.

———. *Tools and Weapons Illustrated by the Egyptian Collection in University College, London, and 2,000 Outlines from Other Sources*. London: British School of Archaeology in Egypt, 1917.

Pettinato, G. *Die Ölwahrsagung bei den Babyloniern*. Band I: *Einleitung*. Studi Semitici 21, 22 (Rome 1966).

Porten, B. *Archives from Elephantine: The Life of an Ancient Jewish Military Colony*. Berkeley: University of California, 1968.

Post, G. E. *Flora of Syria, Palestine, and Sinai.* 2nd ed. 2 vols. Beirut: American, 1932.

Préaux, C. *L'économie royale des Lagides.* Brussels: Fondation Égyptologique, 1939.

———. "Sur les origines des monopoles Lagides." *CE* 29:57 (1954) 316–27.

Puschmann, T. *Alexander von Tralles Original Text und Übersetzung, ein Beitrag zur Geschichte der Medicin.* Vienna, 1878.

Raschke, M. G. "New Studies in Roman Commerce with the East." *Aufstieg und Niedergang der Römischen Welt: Principat.* 9:2. Berlin: Walter de Gruyter, 1978.

Reekmans, T. "The Ptolemaic Copper Inflation." *Studia Hellenistica* 7: *Ptolemaica* (1951) 61–119.

Reil, T. *Beiträge zur Kenntnis des Gewerbes im hellenistischen Ägypten.* Borna-Leipzig: Robert Noske, 1913.

Report on the 1960 World Census of Agriculture, vol. I, part A: Census Results by Countries. Food and Agriculture Organization of the United Nations. Rome, 1966.

Rostovtzeff, M. "Foreign Commerce of Ptolemaic Egypt." *Journal of Economic and Business History* 4:4 (1932) 728–69.

———. "The Foundations of Social and Economic Life in Egypt in Hellenistic Times." *JEA* 6 (1920) 161–78.

———. *A Large Estate in Egypt in the Third Century B.C.: A Study in Economic History.* University of Wisconsin Studies in the Social Sciences and History, 6. Madison, 1922.

———. "Review of N. Lewis, *L'industrie du papyrus dans l'Égypte Grécoromaine.*" *Gnomon* 12 (1936) 51–52.

———. *The Social and Economic History of the Hellenistic World,* 3 vols. Oxford: Clarendon, 1941.

Samuel, A. E. "Illumination by Castor Oil—P.Cornell 1." *BASP* I (1963) 32–38.

Sandy, D. B. "Egyptian Terms for Castor." *CE* 62:123–24 (1987) 49–52.

———. "Oils in Ptolemaic Egypt: The Provisions of *P.Rev.* in Light of the Papyri." Ph.D. diss., Duke University, 1977.

———. "Oil Specification in the Papyri: What Is ΕΛΑΙΟΝ?" *Atti del XVII congresso internazionale di papirologia.* Naples, 1984.

Schnebel, M. *Die Landwirtschaft im hellenistischen Ägypten.* Münchener Beiträge zur Papyrusforschung und antiken Rechtsgeschichte, 7. Munich, 1925.

Segrè, A. *Circolazione monetaria e prezzi nel mondo antico ed in particolare in Egitto.* Roma: Libreria di cultura, 1922.

———. *Metrologia e circolazione monetaria degli Antichi.* Bologna: Nicola Zanichelli, 1928.

______. "Note sull'economia dell'Egitto ellenistico nell' età tolemaica." *Bulletin de la Société d'Archéologie d'Alexandrie* XXIX (1934) 257-305.

Shelton, J. "Artabs and Choenices." *ZPE* 24 (1977) 55-67.

______. "Two Notes on the Artab." *ZPE* 42 (1981) 99-106.

Simari, R., and Martinenghi, G. B. *Olivicoltura e oleificio.* Milan: Ulrico Hoepli, 1950.

Singer, C. "Greek Biology and Its Relation to the Rise of Modern Biology." *Studies in the History and Method of Science.* Vol. II, ed. by Charles Singer. Oxford: Clarendon, 1921.

State Board of Horticulture. *California Olive Industry.* Sacramento: Superintendent, 1900.

Swern, D., ed. *Bailey's Industrial Oil and Fat Products.* Vol. 2, 4th ed. New York: John Wiley, 1982.

Tcherikover, V. "Palestine under the Ptolemies (A Contribution to the Study of the Zenon Papyri)," *Mizraim: Journal of Papyrology, Egyptology, History of Ancient Laws, and Their Relations to the Civilizations of Bible Lands* 4-5 (1937).

Thompson, R. C. *The Assyrian Herbal.* London: Luzac, 1924.

______. *A Dictionary of Assyrian Botany.* London: British Academy, 1949.

Till, W. C. *Die Arzneikunde der Kopten.* Berlin: Akademie, 1951.

Trease, G. E., and Evans, W. C. *Pharmacognosy.* 12th ed. London: Baillière Tindall, 1983.

Turner, E. G. *Greek Papyri: An Introduction.* Oxford: Clarendon, 1980.

U.S. Department of Agriculture. "Agricultural Statistics, 1976." Washington, 1976.

Usher, G. *A Dictionary of Plants Used by Man.* London: Constable, 1974.

Vaughan, J. G. *The Structure and Utilization of Oil Seeds.* London: Chapman and Hall, 1970.

Weiss, E. A. *Castor, Sesame, and Safflower.* New York: Barnes and Noble, 1971.

Westermann, W. L. "Account of Lamp Oil from the Estate of Apollonius." *CPh* XIX (1924) 229-60.

Westermann, W. L., and Laird, A. G. "A Zenon Papyrus at the University of Wisconsin." *JEA* IX (1923) 81-90.

White, K. D. *Agricultural Implements of the Roman World.* Cambridge: Cambridge University, 1967.

______. *A Bibliography of Roman Agriculture,* Bibliographies in Agricultural History 1. University of Reading, Institute of Agricultural History, 1970.

———. *Farm Equipment of the Roman World.* Cambridge: Cambridge University, 1975.

———. *Roman Farming.* Ithaca: Cornell University, 1970.

Wickersham, J. M. "The Financial Prospects of **Ptolemaic Oilmen.**" *BASP* VII (1970) 45–51.

Wilber, D. N. *United Arab Republic: Egypt, Its People, Its Society, Its Culture.* New Haven: HRAF, 1969.

Wilcken, U. *Grundzüge und Chrestomathie der Papyruskunde.* I. Band, *Historischer Teil.* II. Hälfte, *Chrestomathie.* Leipzig: Teubner, 1912.

Willcocks W., and Craig, J. I. *Egyptian Irrigation.* 3rd ed. New York: Spon and Chamberlain, 1913.

Wittfogel, K. A. *Oriental Despotism: A Comparative Study of Total Power.* New Haven: Yale, 1957.

Woenig, F. *Die* ***Pflanzen*** *im alten Aegypten.* Leipzig: Wilhelm Friedrich, 1886.

Youtie, H. C. "Four Short Texts on Papyrus." *ZPE* 29 (1978) 286–92.

———. "Response." *Scriptiunculae Posteriores.* I. Bonn: Rudolf Habelt, 1981.

For additional bibliography not cited above, see p. 82, n. 18 and p. 89, nn. 1–2.

GENERAL INDEX

GREEK WORDS

LATIN WORDS